Raphaël Gobat

Evolution of early-type galaxies at z~1

Raphaël Gobat

Evolution of early-type galaxies at z~1

Differences between field and cluster populations

Südwestdeutscher Verlag für Hochschulschriften

Impressum/Imprint (nur für Deutschland/ only for Germany)
Bibliografische Information der Deutschen Nationalbibliothek: Die Deutsche Nationalbibliothek verzeichnet diese Publikation in der Deutschen Nationalbibliografie; detaillierte bibliografische Daten sind im Internet über http://dnb.d-nb.de abrufbar.
Alle in diesem Buch genannten Marken und Produktnamen unterliegen warenzeichen-, marken- oder patentrechtlichem Schutz bzw. sind Warenzeichen oder eingetragene Warenzeichen der jeweiligen Inhaber. Die Wiedergabe von Marken, Produktnamen, Gebrauchsnamen, Handelsnamen, Warenbezeichnungen u.s.w. in diesem Werk berechtigt auch ohne besondere Kennzeichnung nicht zu der Annahme, dass solche Namen im Sinne der Warenzeichen- und Markenschutzgesetzgebung als frei zu betrachten wären und daher von jedermann benutzt werden dürften.

Verlag: Südwestdeutscher Verlag für Hochschulschriften Aktiengesellschaft & Co. KG
Dudweiler Landstr. 99, 66123 Saarbrücken, Deutschland
Telefon +49 681 37 20 271-1, Telefax +49 681 37 20 271-0, Email: info@svh-verlag.de
Zugl.: München, LMU, Diss., 2009

Herstellung in Deutschland:
Schaltungsdienst Lange o.H.G., Berlin
Books on Demand GmbH, Norderstedt
Reha GmbH, Saarbrücken
Amazon Distribution GmbH, Leipzig
ISBN: 978-3-8381-0619-9

Imprint (only for USA, GB)
Bibliographic information published by the Deutsche Nationalbibliothek: The Deutsche Nationalbibliothek lists this publication in the Deutsche Nationalbibliografie; detailed bibliographic data are available in the Internet at http://dnb.d-nb.de.
Any brand names and product names mentioned in this book are subject to trademark, brand or patent protection and are trademarks or registered trademarks of their respective holders. The use of brand names, product names, common names, trade names, product descriptions etc. even without a particular marking in this works is in no way to be construed to mean that such names may be regarded as unrestricted in respect of trademark and brand protection legislation and could thus be used by anyone.

Publisher:
Südwestdeutscher Verlag für Hochschulschriften Aktiengesellschaft & Co. KG
Dudweiler Landstr. 99, 66123 Saarbrücken, Germany
Phone +49 681 37 20 271-1, Fax +49 681 37 20 271-0, Email: info@svh-verlag.de

Copyright © 2009 by the author and Südwestdeutscher Verlag für Hochschulschriften Aktiengesellschaft & Co. KG and licensors
All rights reserved. Saarbrücken 2009

Printed in the U.S.A.
Printed in the U.K. by (see last page)
ISBN: 978-3-8381-0619-9

Contents

Abstract	xiii
Zusammenfassung	xiv

1 Introduction   1
- 1.1 General properties of early-type galaxies 1
  - 1.1.1 Color-magnitude diagram 3
  - 1.1.2 Fundamental plane 5
- 1.2 Galaxy formation and evolution 5
- 1.3 Thesis outline and goals 7

2 Modeling the spectrophotometric properties of galaxies   13
- 2.1 Model SEDs and spectra 14
  - 2.1.1 Single stellar populations 14
  - 2.1.2 Composite stellar populations 16
- 2.2 Fitting the spectrophotometric data 18
  - 2.2.1 Spectral features and bandpasses 18
  - 2.2.2 Spectrophotometric fitting method 20
- 2.3 Characterizing the star formation history 25
  - 2.3.1 Effects of metallicity 28
- 2.4 Summary 28

3 Robustness of stellar mass estimates   33
- 3.1 The SLACS sample 34
- 3.2 Stellar mass estimates 34
  - 3.2.1 Lensing mass 37
  - 3.2.2 Dynamical mass 38
- 3.3 Photometric stellar mass 40
- 3.4 Comparison between the different stellar mass estimates 41
- 3.5 Visible and dark matter 46
- 3.6 Summary 48

4	**Star formation histories in cluster and field at** $z \sim 1.2$	**53**
	4.1 Data and sample selection	54
	4.2 Spectrophotometric modeling	59
	4.3 Results	60
	4.3.1 Simulations	65
	4.3.2 Considerations on spectral synthesis models	67
	4.3.3 Considerations about metallicity and dust	68
	4.3.4 Rest-frame far-UV flux	69
	4.4 Comparison with semi-analytic models	70
	4.5 Scatter of the red sequence	71
	4.6 Summary	75
5	**Star formation histories in a dense environment at** $z \sim 0.84$	**77**
	5.1 Observations and sample selection	78
	5.1.1 Galaxy colors and luminosities	78
	5.1.2 Stellar mass	81
	5.1.3 Local dark matter density	83
	5.1.4 Projected angular distribution	84
	5.1.5 Composite spectrophotometry	87
	5.2 Stellar population modeling	89
	5.2.1 Star formation histories as a function of luminosity and color	89
	5.2.2 Star formation histories as a function of mass	93
	5.2.3 Star formation histories as a function of environment	95
	5.3 Effect of metallicity and dust	98
	5.4 Summary	103
6	**Moving to higher redshifts: two clusters at** $z \sim 1$ **and** $z \sim 1.4$	**105**
	6.1 The cluster XMMU J1229+0151 at $z = 0.98$	106
	6.1.1 Observations and sample selection	106
	6.1.2 Modeling the star formation history	108
	6.2 The cluster XMMU J2235.3-2557 at $z = 1.39$	115
	6.2.1 Observations and sample selection	117
	6.2.2 Modeling the star formation history	117
	6.3 Summary	125
7	**Summary**	**127**
	Acknowledgements	**143**

List of Figures

1.1	Hubble sequence	3
1.2	Color-magnitude relation in Coma	4
1.3	Projections of the fundamental plane	6
1.4	Models of galaxy formation	8
1.5	Color evolution of stellar population models	9
1.6	The four clusters studied in this thesis	11
2.1	Initial mass functions	16
2.2	Isochrone with stellar library	17
2.3	Example of a composite stellar population model	19
2.4	Passbands and spectral features	21
2.5	Age-metallicity degeneracy	22
2.6	Result of simulated fits to the grid of models	24
2.7	Minimum S/N required for the spectrum	25
2.8	t_{SFR} and t_{fin} of τ-models	26
2.9	Bias of the fit for t_{SFR} and t_{fin}	27
2.10	Bias between the BC03 and M05 models	27
2.11	Metallicity bias for t_{SFR}	30
2.12	Metallicity bias for $T - t_{fin}$	31
3.1	ACS images of SLACS lenses	35
3.2	SED fit of 6 SLACS lenses	42
3.3	Comparison of stellar mass estimates	43
3.4	Comparison of photometric stellar mass estimates	44
3.5	Comparison of M/L ratios	45
3.6	Photometric stellar mass vs total mass	47
3.7	Stellar mass fraction vs R_{Ein}	49
3.8	Stellar mass fraction vs R_{Ein} (binned)	50
3.9	Stellar mass fraction vs z_l (binned)	51
4.1	Composite image of RDCS J1252.9-2927	55
4.2	Stellar mass vs K_s for GOODS and RDCS 1252	56
4.3	Color-magnitude diagram of RDCS J1252.9-2927	57

4.4	Spatial distribution of RDCS 1252 members	58
4.5	Mass completeness of RDCS 1252 and GOODS	59
4.6	Composite spectra of the 10 and 20 brightest ETs of RDCS 1252	61
4.7	Confidence regions of the fit on RDCS 1252 and GOODS	62
4.8	Composite SEDs and spectra of RDCS 1252 and GOODS	62
4.9	t_{fin} of best fitting models to RDCS 1252 and GOODS	63
4.10	t_{SFR} of best fitting models to RDCS 1252 and GOODS	63
4.11	T of best fitting models to RDCS 1252 and GOODS	64
4.12	$m_\star(t)$ of best fitting models to RDCS 1252 and GOODS	64
4.13	Composite spectra of low- and high-mass galaxies	65
4.14	Results of Monte-Carlo simulations on RDCS 1252 and GOODS	66
4.15	t_{SFR} and t_{fin} of best fitting M05 models to RDCS 1252 and GOODS	67
4.16	$m_\star(t)$ for different metallicities	68
4.17	Best fit E(B-V) to the SEDs of RDCS 1252 and GOODS	69
4.18	U-band flux of best fit models to RDCS 1252 and GOODS	70
4.19	Observed t_{SFR} vs predictions of semi-analytic models	72
4.20	Observed $m_\star(t)$ vs predictions of semi-analytic models	72
4.21	Color-magnitude relation of best fit models to RDCS 1252	73
4.22	Predicted red sequence scatters	74
4.23	Median star formation history of RDCS 1252 and GOODS	75
5.1	ACS image of RX J0152.7-1357	79
5.2	Photometric filters used for RX J0152.7-1357	81
5.3	Color-magnitude diagram of RX J0152.7-1357	82
5.4	Red sequence bins in RX J0152.7-1357	82
5.5	Comparison of photometric stellar mass estimates of ETGs	83
5.6	Distribution of $M^\star_{phot}$ of ETGs in RX J0152.7-1357	84
5.7	Dark matter density regions in RX J0152.7-1357	85
5.8	Angular position regions in RX J0152.7-1357	86
5.9	Composite spectra of the red sequence bins in RX J0152.7-1357	87
5.10	Composite spectra of the mass bins, dark matter and radial regions	88
5.11	t_{SFR} and t_{fin} for the red sequence bins	90
5.12	Position of the red sequence galaxies of RX J0152.7-1357	92
5.13	$D_n(4000)$ vs Hδ_A and H6 for the red sequence bins	93
5.14	t_{SFR} and t_{fin} for the mass selected bins	94
5.15	t_{SFR} vs $M^\star_{phot}$ for the red sequence bins	94
5.16	t_{SFR} and t_{fin} for the dark-matter density regions	95
5.17	t_{SFR} and t_{fin} for the angular position regions	96
5.18	t_{SFR} vs $M^\star_{phot}$ as a function of environment	98
5.19	Positions of ETGs in RX J0152.7-1357 sorted by mass	99
5.20	E(B-V) vs SFR(OII)	101
5.21	$D_n(4000)$ vs Hδ and H6 of red sequence galaxies	102
5.22	Median star formation histories of ETGs in RX J0152.7-1357	103

List of figures

6.1 ACS image of XMMU J1229+0151 107
6.2 Spatial distribution of XMMU J1229+0151 galaxies 109
6.3 Color-magnitude diagram of XMMU J1229+0151 111
6.4 Spectra of passive galaxies in XMMU J1229+0151 112
6.5 SED fit of a XMMU J1229+0151 member 113
6.6 Distribution of $M^\star_{phot}$ of XMMU J1229+0151 ETGs 113
6.7 Confidence regions of the fit to the XMM 1229 sample 114
6.8 Composite SED and spectrum of XMM 1229 with best fit models 114
6.9 t_{SFR} and t_{fin} of best fit models to the XMM 1229 sample 115
6.10 t_{SFR} vs $M^\star_{phot}$ of red sequence galaxies in XMMU J1229+0151 116
6.11 Composite image of XMMU J2235.3-2557 118
6.12 Spatial distribution of galaxies in XMMU J2235.3-2557 119
6.13 Distribution of $M^\star_{phot}$ of ETGs in XMMU J2235.3-2557 120
6.14 Results of the fit to the XMM 2235 sample 122
6.15 Composite spectrum of the XMM 2235 sample 123
6.16 Composite spectrum of the core and periphery of the XMM 2235 sample . 123
6.17 $(U-V)_z$ of the cluster samples 124
6.18 Median star formation histories of the four cluster samples 125

7.1 Star formation histories of cluster and field ETGs 128
7.2 Star formation histories in RX J0152.7-1357 129
7.3 High-z clusters and timeline of ETG formation 132

List of Tables

3.1	Properties of the SLACS lenses	36
3.2	Properties of the SLACS II lenses	36
4.1	Cumulative spectroscopic completeness of RDCS 1252 and GOODS	59
4.2	Results of the fit to RDCS 1252 and GOODS	63
5.1	Definition of bins in RX J0152.7-1357	80
5.2	Results of the fit to the different bins of RX J0152.7-1357	97
6.1	Properties of the passive spectroscopic members of XMMU J1229+0151	110
6.2	Results of the fit to the XMM 1229 sample	114
6.3	Properties of the passive spectroscopic members of XMMU J2235.3-2257	117

Abstract

In this thesis, we have studied several aspects of the evolution of high-redshift ($0.8 < z < 1.4$) early-type galaxies across a range of environments, by modeling their stellar population properties, thus inferring their star formation histories. For this purpose, we have used an exceptional dataset, in terms of quality, depth and wavelength coverage, combining spectrophotometric observations from the VLT, HST and Spitzer telescopes, which is hardly matched by other investigations. We have developed a novel method which combines both the SED and spectra of galaxies to model the underlying stellar populations with spectral synthesis models.

We have checked the robustness of stellar mass estimates obtained from SED modeling by comparing them to stellar mass estimates from the literature based on a combination of strong lensing and stellar dynamics. We have found that those two independent estimates are in excellent agreement. We have found that the stellar mass of the galaxies is proportional to the total mass and that the fraction of stellar to dark matter is constant out to one effective radius, implying that the profiles of the dark and stellar matter distributions in these galaxies are similar.

We have then compared the star formation histories of field and cluster galaxies at $z \approx 1.2$, a crucial test for galaxy formation models, and found a difference of ~ 0.5 Gyr in their respective star formation timescales, with field galaxies having longer star formation histories. This difference is much smaller than that observed at low redshift and implies that $\sim 10\%$ of the stellar mass of early-type galaxies was assembled at $z < 1$. By modeling the photometric properties of red sequence galaxies in a $z = 1.24$ cluster, we conclude that the tight red sequence observed in this cluster was established over ~ 1 Gyr, starting at $z \sim 2$.

We have studied the star formation histories in a massive $z = 0.84$ cluster and found a strong dependence of galaxy age with mass and clustercentric distance, which can not be attributed to metallicity differences. The massive core galaxies formed at $z > 3$ and became passive at $z \gtrsim 2$, while early-type galaxies in the cluster outskirts are ~ 1.5 Gyr younger. In particular, we found a population of post-starburst galaxies at the edge of the cluster which occupies the faint blue end of the cluster red sequence.

We have also studied the star formation histories of two other high-redshift clusters, at $z = 0.98$ and $z = 1.39$. The analysis of the latter, the most distant massive cluster known to date, has revealed an already old galaxy population, with signs of strong radial gradients, suggesting that we are indeed approaching the formation epoch of the red cluster population. We have compared the slopes of the age-mass relations of early-type galaxies in the four clusters and found them to be nearly identical. We have also found a variance of ~ 0.5 Gyr in the formation epochs of massive early-type galaxies between clusters, which provides an interesting test for models of galaxy formation and evolution.

Zusammenfassung

Gegenstand der vorliegenden Arbeit war die Untersuchung von verschiedenen Aspekten der Entwicklung von Frütyp-Galaxien bei hohen Rotverschiebungen ($0.8 < z < 1.4$) und einer Reihe von Umgebungen. Der dabei verwendete Datensatz war bezüglich Qualität, Tiefe und Wellenlänge Abdeckung ausserordentlich hoch und konnte erreicht werden durch die Kombination von spektrophotometrischen Beobachtungen mit den VLT, HST und Spitzer Teleskopen. Dank einer von uns für den Zweck entwickelten neuartigen Methode, dessen Novum und Nutzen darin besteht, die SED sowie die Spektren der Galaxien zu kombinieren, konnten wir ein Modell für die zugrunde liegende stellare Bevölkerung herzustellen, mit Hilfe von Spektralsynthese-Modelle.

Wir haben zuerst die Robustheit von stellaren Massebestimmungen aus der SED-Modellierung überprüft, durch die Gegenüberstellung mit jenen aus der Literatur, auf der Grundlage den starken Linseneffekts und der stellaren Dynamik. Wir haben festgestellt, dass die Ergebnisse dieser beider unabhängiger Bestimmungen in sehr guter Uebereinstimmung zu einander stehen und dass sich die stellare Masse der Galaxien proportional zu ihrer gesamten Masse verhält. Der Anteil von stellarer zu dunkler Materie bleibt konstant bis zum wirksamen Radius, was bedeutet, dass stellare und dunkle Materie ähnlich verteilt sind.

Darüber hinaus haben wir die Geschichten der Sternenentstehung im Feld und in den Haufen Frütyp-Galaxien miteinander verglichen. Dies ist ein wichtiger Test für Modelle der Galaxien Entstehung. Dabei haben wir eine Differenz von ~ 0.5 Gyr in ihren jeweiligen Sternbildungszeiten gefunden. Dieser Unterschied ist viel kleiner als der, der bei niedrigen Rotverschiebungen beobachtet wurde, und bedeutet, dass $\sim 10\%$ der stellaren Masse der Frütyp-Galaxien bei $z < 1$ gebildet wurde. Durch die Modellierung der photometrischen Eigenschaften von Rot-Sequenz-Galaxien in einem $z = 1.24$ Galaxienhaufen haben wir den Schluss gezogen, dass die rote Sequenz dieses Haufens über ~ 1 Gyr zusammengesetzt wurde, beginnend bei $z \sim 2$.

Wir haben die Geschichten der Sternenentstehung in einem massivem $z = 0.84$ Galaxienhaufen analysiert und dabei eine starke Abhängigkeit des Galaxienalters von der stellaren Masse und von der Entfernung vom Haufenzentrum festgestellt. Die massiven Kerngalaxien bildeten sich bei $z > 3$, um sich dann nach $z \gtrsim 2$ passiv entwickelt, während die Frütyp-Galaxien am Haufenrand ~ 1.5 Gyr jünger sind. Bemerkenswert war die Entdeckung von einer Bevölkerung von Post-Starburst-Galaxien am Rand vom Haufen.

Zusäztlich haben wir die Geschichten der Sternenensetehung in zwei anderen Haufen untersucht, und zwar bei $z = 0.98$ und $z = 1.39$. Im Letzteren, bekannt als der am weitesten entfernt massiven Haufen, fanden wir eine alte Galaxien-Bevölkerung sowie einen starken radialen Altersgradient, was darauf hindeutet, dass wir uns der Entwicklungsepoche der roten Haufen-Galaxien nähern. Schliesslich stellten wir auch fest, dass es eine Varianz von ~ 0.5 Gyr in der Entwicklungsepoche der massiven Frütyp-Haufengalaxien gibt, die einen interessanten Test für Modelle der Galaxienentstehung darstellt.

Chapter 1

Introduction

The problem of the formation and evolution of galaxies is still one of the great questions in cosmology, more than eighty years after galaxies were recognized for what they are, "island universes" of their own rather than mere nebulae in ours. As observations went deeper and better tools became available, increasingly sophisticated galaxy formation models were formulated. However, even the most recent, state-of-the art hierarchical merging models, which are rooted in the dark matter paradigm, fail to reproduce all the observable properties of galaxies. Progress had been slow initially, in part because the vast range of physical processes, both internal and external to the galaxies, that influence their formation and evolution required a comprehensive picture of the cosmos and the understanding of such diverse subjects as dark matter, active galactic nuclei and stellar evolution. Also, a statistical analysis of galaxy properties that can effectively constrain models of formation and evolution requires vast quantities of data across as many cosmic epochs as possible. Unfortunately, in the nearby Universe the traces of the early formation of galaxies have been smoothed out by the many billion years of evolution elapsed since then, hence the interest in observing galaxies at high redshift, close to their epoch of formation. This was for a long time severely limited by the power of the instruments but in the last ten years the situation has improved tremendously, as several $z > 1$ galaxy clusters have been confirmed and deep multiband, sometimes multi-observatory, galaxy surveys have been carried out. In this work, we made use of some of these deep datasets to constrain star formation histories, and thus possible evolution processes, of high redshift galaxies. We concentrated on early-type galaxies, a class well suited to this purpose. Let us briefly recapitulate their most important properties.

1.1 General properties of early-type galaxies

When Edwin Hubble first proposed his famous morphological classification scheme (Hubble [1926]), he divided galaxies into three classes: the ellipticals (E), spirals (S) and irregulars. A transition class between the elliptical and spiral galaxies, dubbed "lenticulars" (or S0), was also postulated. Because of their positions in Hubble's sequence (Fig. 1.1), and possibly for historical reasons too (e.g. Jeans [1919]), he called the ellipticals and lenticulars "early-type" galaxies (or ETGs) and the spirals "late-type". Ellipticals galaxies are designated as En, with $n = 10(1 - b/a)$, b/a being the apparent axial ratio, ranging from 0 for spheroidals to 7 for the most oblate elliptical galaxies. Elliptical galaxies have smooth and mostly featureless surface brightness profiles which can be approximated by the empirical de Vaucouleurs law ([1948]):

$$I(R) = I_e\ exp\Big(-7.67\Big(\frac{R}{R_e}\Big)^{1/4}\Big) \quad (1.1)$$

where R is the projected distance from the center and R_e and I_e the half-luminosity radius and surface brightness respectively. While elliptical galaxies cover a large range of scales and masses, due to their distance the ellipticals considered here are all in the high mass range, from 10^{10} to $10^{12} M_\odot$. Kinematically, elliptical galaxies show little to no total rotation and large velocity dispersions, typically between 100 and 400 kms^{-1}. S0 galaxies, on the other hand, are characterized by two components, a central bulge, which is similar to an elliptical, and a disk whose brightness profile can be described by a Sérsic profile

$$I(R) = I_e\ exp\Big\{-b_n\Big[\Big(\frac{R}{R_e}\Big)^{1/n}-1\Big]\Big\} \quad (1.2)$$

of index $n = 1$. Together, early-type galaxies represent $\sim 17\%$ of the total number of galaxies in the nearby Universe but $\sim 57\%$ of the total mass (e.g. Baldry et al. [2004], Renzini et al. [2006]). While E and S0 galaxies are structurally different, they have otherwise similar properties:

- they are passively evolving stellar systems. An early-type galaxy shows little to no ongoing star formation, and this since at least ~ 1 Gyr (e.g. Sandage & Visvanathan [1978]). In addition, early-type galaxies are dominated by old stellar populations. Elliptical galaxies in clusters are understood to have formed at redshifts higher than 2 (e.g. Bernardi et al. [1998], van Dokkum et al. [2001a]), with field early-type galaxies being ~ 1 Gyr younger.

- consequently, there are no emission line regions and while appreciable amounts of dust can be seen in early-type galaxies, in the form of dust lanes or disks, the dust structures do not affect significantly their integrated optical and near-infrared emission. Combined with the lack of ongoing star formation, this makes the study of their star formation history possible, using spectroscopy and broad-band photometry, whereas the integrated light of late-type galaxies is dominated by the ongoing star formation.

- finally, early-type galaxies trace the highest peaks of matter density in the Universe. While early-type galaxies are found in all environments, they constitute a high fraction of the cluster galaxy population, up to $\sim 80\%$ in the cores of nearby clusters (e.g. Dressler [1980]). This makes early-type galaxies well suited to test the different models of structure formation.

1.1 General properties of early-type galaxies

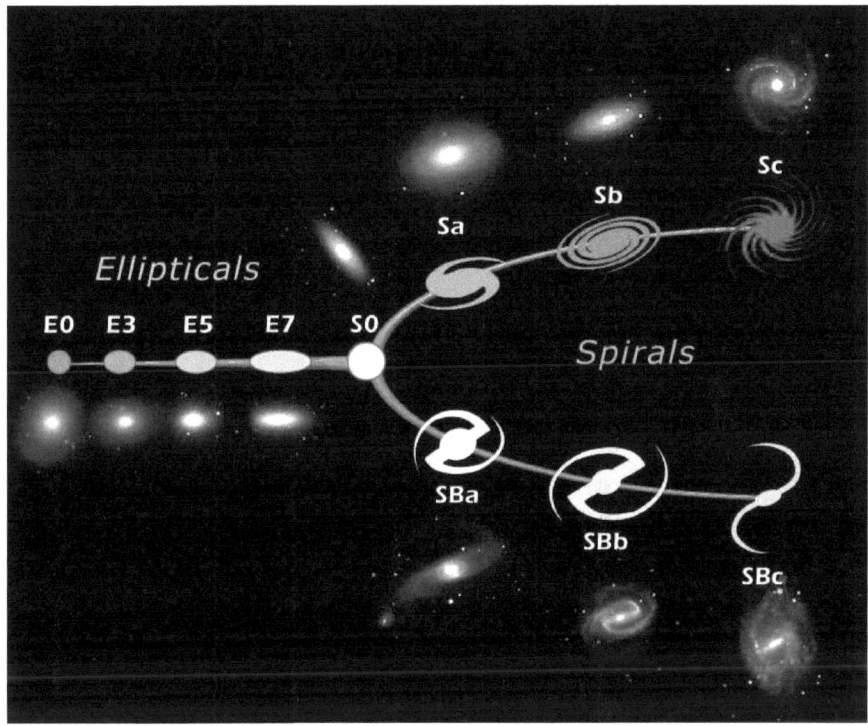

Figure 1.1: Hubble "tuning-fork" sequence showing, from left to right, elliptical galaxies (E), S0 galaxies and spiral galaxies, both normal (S) and barred (SB). Credit: STScI

In addition, several scaling relations have been found between the various photometric and kinematic properties of early-type galaxies.

1.1.1 Color-magnitude diagram

The color magnitude relation is the first and oldest of the scaling relations (Baum [1959]). It correlates stellar population properties of galaxies, expressed in their colors, with their mass, traced by the galaxy luminosity. In color-magnitude space, early-type galaxies correlate tightly in a so-called "red sequence" while most late-type galaxies are more loosely distributed in what is sometimes named the "blue cloud". The red sequence is characterized by two parameters, its scatter and its slope. The small scatter of the red sequence (e.g. $\delta(U - V) = 0.05$ for Coma; Eisenhardt [2007]) found in all clusters requires that the star formation histories of early-type cluster galaxies be well synchronized. This in turn implies that either all early-type cluster galaxies formed at high redshift or that later

1. Introduction

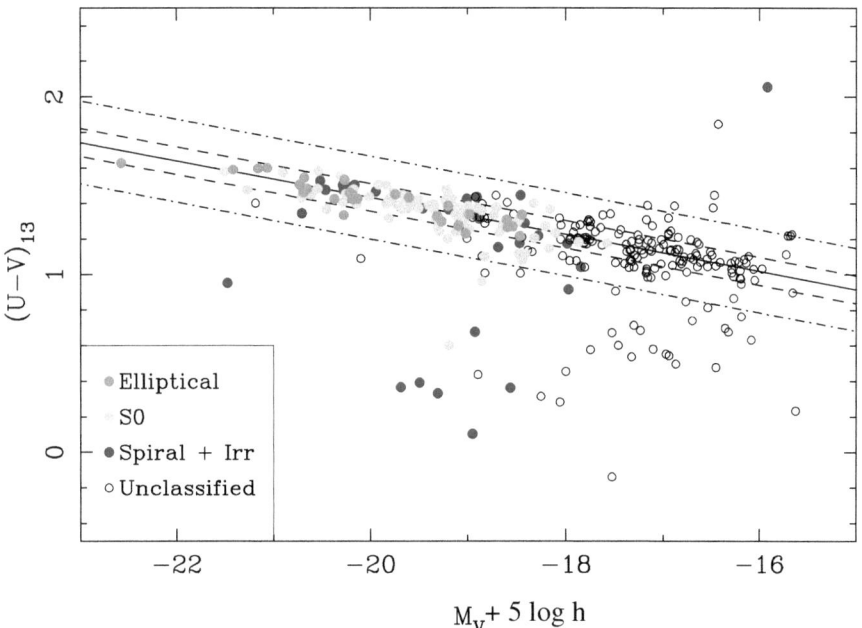

Figure 1.2: $(U-V) - M_V$ color-magnitude relation of galaxies in the Coma cluster. Elliptical galaxies are shown in red, S0 galaxies in green and late-type galaxies in blue. The solid line shows the best fit linear correlation while the dashed and dot-dashed lines show the 1σ and 3σ scatters respectively. Credit: Bower et al. ([1999]).

bursts of star formation account for less than 10% of the total stellar mass present today (e.g. Bower, Lucey & Ellis [1992]). The slope of the red sequence was first interpreted as an effect of increasing metal content with luminosity (Faber [1973]), as less massive galaxies, due to a weaker potential well, are less able to retain the enriched gas dispersed by supernova winds (Arimoto & Yoshi [1987]). In principle, however, the color-magnitude relation could instead be an effect of increasing age with luminosity, the former affecting color in the same way as metallicity. In this case, the slope of the red sequence should change quickly at high redshift but, while the zeropoint of the color-magnitude relation does indeed change with redshift, little to no evolution of the slope has been observed up to $z \sim 1.3$ (e.g. van Dokkum et al. [2000], Mei et al. [2006b]). Together, the relatively small scatter and constant slope of the red sequence up to $z \sim 1.4$ puts strong constraints on galaxy formation models.

1.1.2 Fundamental plane

The second scaling law, called the Faber-Jackson relation (Faber & Jackson [1976]), correlates the total luminosity L and the central velocity dispersion σ of early-type galaxies:

$$L \propto \sigma^\gamma \qquad (1.3)$$

with γ being very close to 4. This implies that, for a virialized galaxy, the mass-to-light ratio M/L and the surface brightness $I = L/4\pi R^2$ are constant, as $\sigma^2 \propto GM/R$. This relation provides a very useful tool to measure cosmological distances. The third scaling relation, called the Kormendy relation (Kormendy [1977]), relates the effective radius R_e of early-type galaxies to their mean surface brightness $\langle I \rangle_e$:

$$\langle I \rangle_e \propto \log R_e \qquad (1.4)$$

The Kormendy relation has been used to study the structural properties of early-type galaxies and their evolution up to $z > 1$ (e.g. Fasano et al, [1998], Ziegler et al. [1999], Holden et al. [2006], Scarlata et al. [2007]). The Faber-Jackson and Kormendy relations imply that, in the $\{R_e, \sigma, I_e\}$ space, early-type galaxies are distributed along a plane (see Fig. 1.3), called the "fundamental plane" (Djorgovski & Davis [1987], Dressler et al. [1987]) and described by the equation

$$\log R_e = \alpha \log \sigma + \beta \log \langle I \rangle_e + \gamma \qquad (1.5)$$

where α, β and γ depend on the bandpass used for measuring the luminosity. On the other hand, the virial theorem requires that galaxies must satisfy

$$\log R_e \sim 2 \log \sigma - \log \langle I \rangle_e + C \qquad (1.6)$$

assuming a constant M/L and the structural homology of early-type galaxies (i.e. that they have the same mass distribution and velocity dispersion profiles). The significant discrepancy between the coefficients of the fundamental plane and the prediction of the virial theorem (e.g. Jørgensen, Franx & Kjærgaard [1996]) imply that either the virial hypothesis is not valid, that early-type galaxies are not homologous (e.g. Graham, Trujillo & Caon [2001]) or that the M/L changes with galaxy parameters, typically luminosity. In addition, the M/L of early-type galaxies has been shown to evolve with redshift (e.g. Holden et al. [2006]), consistently with a high formation redshift and subsequent passive evolution.

1.2 Galaxy formation and evolution

The observed properties of galaxies have traditionally been interpreted in the framework of two different classes of galaxy formation scenarios (see Fig. 1.4): the monolithic collapse model (e.g. Eggen, Lynden-Bell & Sandage [1962], Larson [1975], Rees & Ostriker [1977]) and the hierarchical formation model (e.g. Toomre [1977], White & Rees [1978]). In the

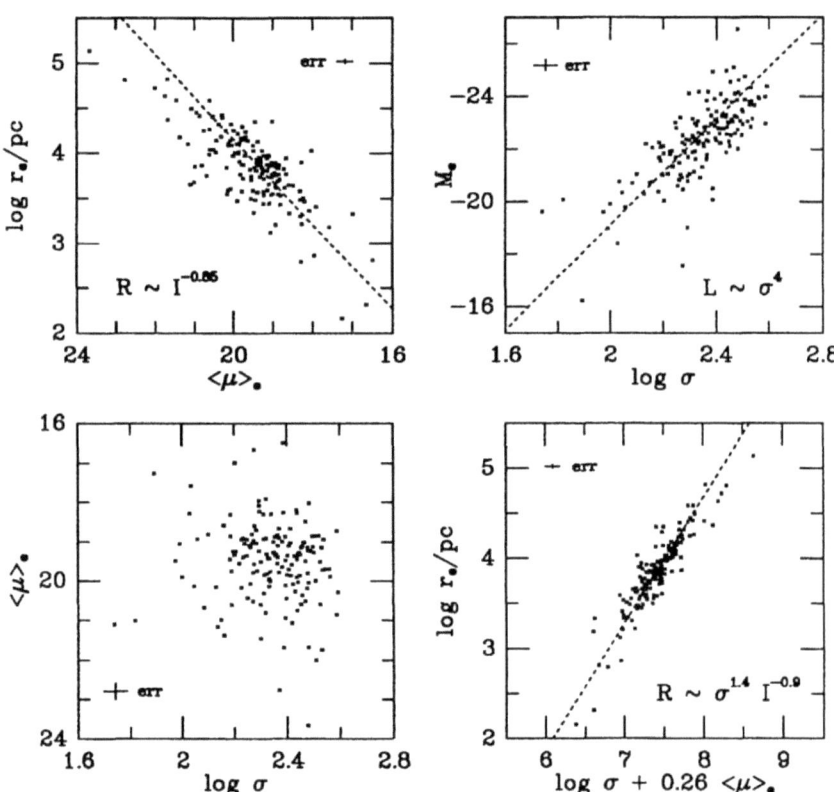

Figure 1.3: Projections of the fundamental plane: the relation between the radius and mean surface brightness (top left), the Faber-Jackson ([1976]) relation between luminosity and velocity dispersion (top right), the relation between mean surface brightness and velocity dispersion (i.e. the fundamental plane seen almost face-on) and the relation between the radius and a combination of surface brightness and velocity dispersion (i.e. the fundamental plane seen edge-on). Credit: Djorgovski & Davis ([1987]), in Kormendy & Djorgovski ([1989]).

former, galaxies formed at high redshift in a single event through gravitational collapse. Early-type galaxies would then cease star formation shortly afterwards, then evolve passively, and a spheroid might later accrete a disk if enough gas is present in its environment. In the hierarchical scenario, massive galaxies form through the merging of smaller units, elliptical galaxies forming from the disruption of disks during major merging events. Thus, the monolithic collapse model has massive galaxies being in place very early on while hierarchical formation implies that less massive galaxies formed first. While the hierarchical merging scenario arises naturally from cold dark matter (CDM) models (e.g. White & Rees [1978], Davis et al. [1985]), the high formation redshifts found for massive ellipticals (see above) tend to support a single-event formation scenario and hierarchical models of galaxy formation have generally struggled to predict the uniformly high ages deduced for early-type galaxies and their number fraction (e.g. McCarthy et al. [2004b], Daddi et al. [2005]; see also Chapter 4). Another seemingly "anti-hierarchical" behavior exhibited by early-type galaxies is the fact that their ages correlate with mass, i.e. that more massive galaxies appear to have formed over a shorter time span (e.g. Thomas et al. [2005], Treu et al. [2005]). On the other hand, the hierarchical formation model predicts that galaxies in high density environments form earlier than their counterparts in lower density regions (e.g. De Lucia et al. [2006]). This has been effectively confirmed by various studies at low and high redshift (e.g. Thomas et al. [2005], Sánchez-Blázquez et al. [2006], Clemens et al. [2006], van Dokkum & van der Marel [2007]; also, Chapter 4). Indeed, the fact that properties of galaxies such as age and morphology (e.g. Spitzer & Baade [1951], Dressler et al. [1980]) depend on the environment provides a strong argument in favor of the hierarchical formation scenario. For this reason, modern models of galaxy formation and evolution are based on this latter scenario. They typically follow the collapse and merging of dark matter haloes of proto-galaxies, computed either analytically using the extended Press-Schechter formalism (e.g. Lacey & Cole [1993]) or from N-body simulations. The formation of the galaxies themselves is then modeled by adopting a treatment of the baryons associated with a given halo where star formation is regulated by internal processes, such as the cooling rate of the gas and feedback by supernovae, as well as interactions between the dark matter haloes. The determination of the cosmological model and precise measurements of its parameters (e.g. Spergel et al. [2007]) makes it possible to constrain the hierarchical scenario of galaxy formation using the observed properties of galaxies. As the star formation history of galaxies in a hierarchical merging model depends in part on the merging history of the dark matter haloes, early-type galaxies, which have evolved in a mostly passive way since their last significant episode of star formation at high redshift, appear ideally suited to test such models.

1.3 Thesis outline and goals

In this work, we used multiwavelength observations of four of the highest redshift clusters, from $z \sim 0.8$ to $z \sim 1.4$, and of two field samples to constrain the star formation histories of galaxies at these redshifts. We focused on early-type galaxies because their

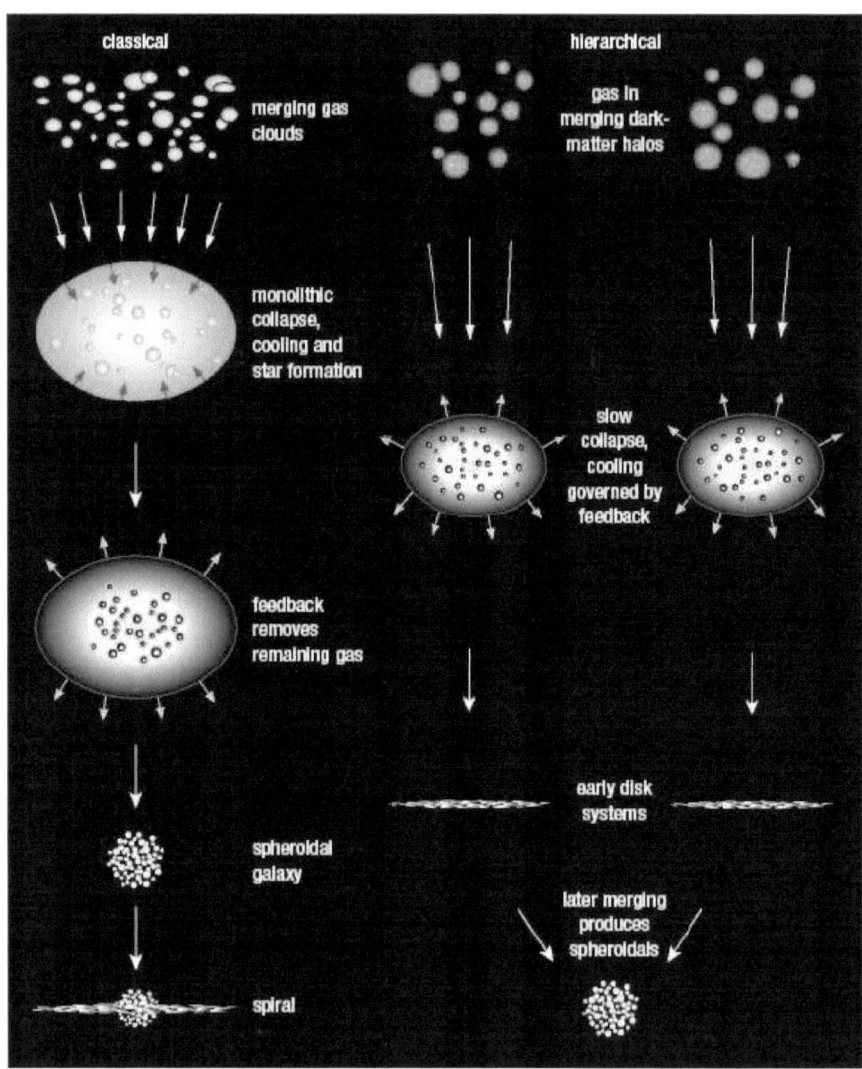

Figure 1.4: Monolithic model of galaxy formation (left), where galaxies form in isolation and their evolution is dependent mostly on initial conditions, and the hierarchical model (right), where galaxies form and evolve through successive mergers of smaller haloes. In this scenario, the evolution of galaxies is much more dependent on environment. Credit: Ellis et al. ([2000]).

1.3 Thesis outline and goals

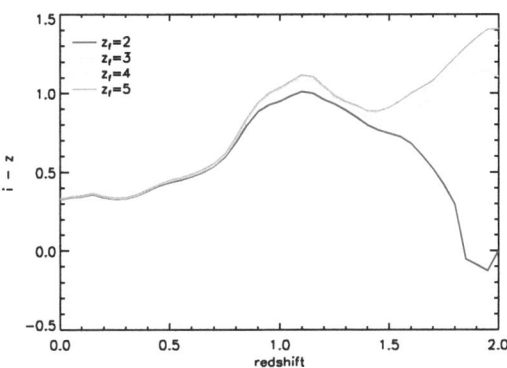

Figure 1.5: $i - z$ color evolution of Kodama & Arimoto ([1997]) model stellar populations from $z = 2$ to the present, for galaxies with formation redshifts of 2, 3, 4 and 5. The colors at low redshift are very similar but start to increasingly diverge at $z > 1$.

ideal stellar population and structural characteristics, as detailed above. In this case high redshift observations are of particular interest, as the star formation histories of early-type galaxies can be more easily retraced when they are observed near their epoch of formation, rather than at low redshift when their photometric and spectroscopic properties have been homogenized by several billion years of passive evolution (see Fig. 1.5). Likewise, galaxy clusters being the most biased regions in the Universe, we can expect any effect of environment on the evolution of galaxies to be much more pronounced in clusters than in low density regions. Lastly, galaxy clusters make it possible to study a sample of coeval galaxies and the variation of their stellar population properties across a range of environmental densities and intrinsic parameters (such as luminosity and morphological type), without having to worry about cosmic variance or the biases that are inevitably introduced when comparing galaxies spread over a large swath of cosmic time.

This thesis is structured as follows:
In Chapter 2, we describe stellar population synthesis models and the novel method used to compare them to the observed data. We quantified the biases due to the method itself, the choice of models and of model parameters, as well as the constraints set by the data.
In Chapter 3, we compare stellar mass estimates obtained from broadband photometry to recently published stellar mass estimates from gravitational lensing and stellar dynamics for a sample of elliptical galaxies acting as gravitational lenses. We also discuss the implications for the dark matter distribution in these galaxies. The results of this analysis have been partly published in Grillo et al. ([2008]).
In Chapter 4, we investigate differences in the star formation histories of massive early-type galaxies at $z \sim 1.2$ in a low density environment and the galaxy cluster RDCS J1252.9-

2927. We also compare the inferred star formation histories with the predictions of a model of galaxy formation and evolution. The results in this Chapter were partly published in Gobat et al. ([2008]) and Menci et al. ([2008]).

In Chapter 5, we present an analysis of the stellar population parameters of early-type galaxies in the massive galaxy cluster RX J0152.7-1357, at $z \sim 0.8$, and their variation with the intrinsic properties of galaxies and the local environment. Some results from this Chapter will be published in Demarco et al. ([2009]).

Finally, in Chapter 6 we extend our analysis to two other clusters, XMMU J1229+0151 and XMMU J2235.3-2557, at $z \sim 1$ and $z \sim 1.4$ respectively. Results from this Chapter will be partly published in Santos et al. ([2009]) and Rosati et al. ([2009]).

Throughout this work, we assumed a ΛCDM cosmology with $\Omega_m = 0.3$, $\Omega_\Lambda = 0.7$ and $H_0 = 70$ kms^{-1} Mpc^{-1}. All magnitudes are given in the AB system (Oke [1974]) unless stated otherwise.

1.3 Thesis outline and goals

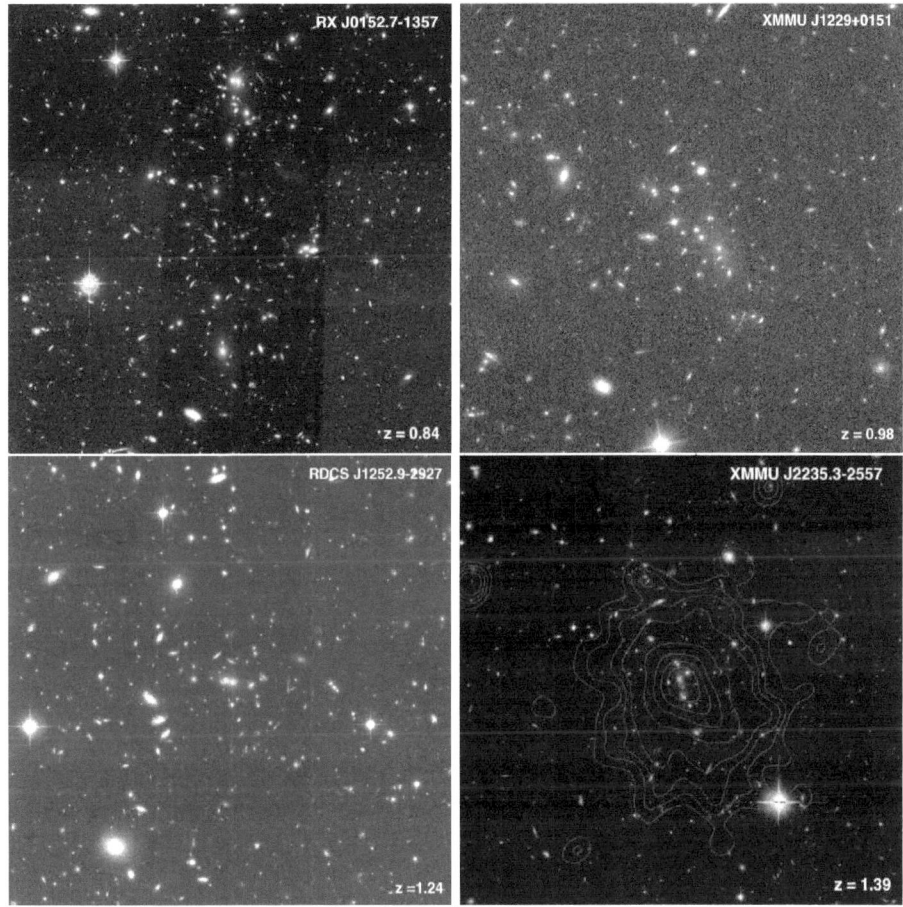

Figure 1.6: Color composite images of the four clusters studied in this thesis: RX J0152.7-1357 at $z = 0.84$ (top left), XMMU J1229+0151 at $z = 0.98$ (top right), RDCS J1252.9-2927 at $z = 1.24$ (bottom left) and XMMU J2235.3-2557 (bottom right).

Chapter 2

Modeling the spectrophotometric properties of galaxies

There are many ways in which useful information can be coaxed out of photometric data. A single band image already contains information on the state and structure of the target galaxies. Luminosity functions only require a single band and allow one to trace some aspects of the evolution of entire galaxy populations (e.g. De Propris et al. [2007]). With two judiciously chosen filters, a color magnitude diagram can be made that will make possible the rough characterization of the stellar populations of the observed galaxies. With more bandpasses, different phases of star formation in the galaxy can be distinguished and its star formation history begins to be revealed. On the other hand, spectra pack a lot of information from the get go but are harder to obtain. Combined with the fact that they are often taken with the intent of deriving accurate redshifts, and as such are just good enough for this purpose, much of the useful spectral information can be buried in noise. Also, as several effects can influence the same spectral feature, their interpretation can be delicate. It is therefore convenient to combine spectroscopy and photometry for maximal explaining power.

While population of galaxies can be compared based on their observational data only (e.g. Dressler et al. [2004], Luo et al. [2007]), without additional reference points, stellar population synthesis models offer a powerful tool for interpreting the photometry and spectroscopy of galaxies and deriving key stellar population parameters. They allow for a deeper study of the variation of stellar populations among galaxy samples and the tracing of star formation histories. Here we propose to derive stellar population parameters from the combined photometry and spectroscopy of early-type galaxies with the use of of stellar population models.

This Chapter is organized as follows. In Section 2.1, we describe the stellar population synthesis models used in this work to fit the observed data, as well as the chosen star formation history. In Section 2.2, we describe briefly the relevant spectral features that are compared to the models' and detail our method to fit both the SED and spectrum of galaxies. In Section 2.3, we define the two parameters we used to characterize the star formation histories of early-type galaxies.

2.1 Model SEDs and spectra

2.1.1 Single stellar populations

We computed our stellar population models using so-called single stellar population (SSP) templates. Each template is the composite spectrum of a stellar population of a given age, where all the stars in said population have been born at the same time. In this work, we considered two sets of SSP templates, those of Bruzual & Charlot ([2003], hereafter BC03) and those of Maraston ([2005], hereafter M05). Here we recapitulate the characteristics of each set of templates and the main ingredients for computing a single stellar population spectrum:

First, one needs a **library of stellar spectra**, which samples as much as possible of the space of stellar parameters (spectral type, luminosity class, metallicity, abundance of α-elements, etc.) and can be made up from theoretical or observed spectra. The advantage of a theoretical library is that the parameter space can be sampled finely and the metallicity and α-abundance set precisely. However, a theoretical library relies on model atmospheres which are limited by our knowledge of the physical processes leading to the spectral features in the stellar continuum. This limits the resolution of theoretical spectra and also means that they might not reproduce accurately enough the spectra of actual stars. On the other hand, an empirical library does not suffer from the same problems, as it does not depend on a (necessarily incomplete) list of spectral lines. However, the wavelength and parameter space coverage of an empirical library is limited by the available observatories and the observable stellar populations (the solar neighborhood, the Galactic bulge and the Magellanic Clouds) respectively. In addition, the stellar parameters (such as effective temperature and metallicity) of the stars are often obtained from a number of sources, using different methods and models, which reduces the overall consistency of the library. Both the BC03 and M05 models use the library of theoretical spectra of Lejeune, Cuisinier & Buser ([1997], [1998]; hereafter BaSeL) updated by Westera et al. ([2002]). This library provides theoretical spectra of stars at a resolution of 10 to 20 Å FWHM in the wavelength range from 91 Å to 160 μm and for metallicities from Z=0.0001 to Z=0.1. At solar metallicity and in the range 1150 Å to 2.5μm, the BC03 models can make use of the Pickles ([1998]) library of galactic stellar spectra, with a resolution of 5 Å. Furthermore, in the wavelength range from 3200 to 9500 Å, the BC03 models use the high resolution observational library of Le Borgne et al. ([2003], hereafter STELIB) at 3 Å resolution and sampled at 1 Å pixel^{-1} (to which the Pickles library is also resampled).

Next, one needs a set of tracks, in the theoretical ($\log L$ vs $\log T_{eff}$) Herzsprung-Russel (HR) diagram, that describe the evolution of all stars in the given parameter space. For given abundance ratios and metallicity, interpolating over all the tracks at time T yields

2.1 Model SEDs and spectra

an **isochrone** of age T. The BC03 models offer the choice of three sets of stellar evolution tracks. Here we used those computed by Alongi et al. ([1993]), Bressan et al. ([1993]), Fagotto et al. ([1994a], [1994b]) and Girardi et al. ([1996]). The tracks extend from the zero-age main sequence until the thermally pulsing asymptotic giant branch regime (TP-AGB for short) or core carbon ignition, depending on the stellar mass, and are commonly referred to as the "Padova 1994" library. While a newer, revised library exists (Girardi et al. [2000]), it produces worse agreement with galaxy colors (Bruzual & Charlot [2003]) than the Padova 1994 library. For the main-sequence phase of stellar evolution, the M05 models use stellar evolution tracks from Cassisi et al. ([1997a], [1997b], [2000]).

Where the BC03 and M05 models differ significantly is in their treatment of the post-main sequence phases of stellar evolution. The BC03 models use the Padova 1994 tracks up to the beginning TP-AGB phase and the tracks of Vassiliadis & Wood ([1993]) for the asymptotic giant branch. The M05 models, on the other hand, estimate the contribution of post-main sequence stars using the fuel consumption theorem (Renzini & Buzzoni, [1986]). This results in a large difference in the near-IR between the two prescriptions (Maraston [2005]) in the age range 0.2 to 2 Gyr, when the contribution of TP-AGB stars is maximal (e.g. Frogel et al. [1990]).

Finally, the proportions of stars of different initial masses are given by the **initial mass function** (IMF) $\phi(m)$, defined such that $\phi(m)dm$ is the number of stars born with masses between m and $m + dm$. Both models offer the choice of a Salpeter ([1955]) IMF, which is a simple power-law ($\phi(m) \propto m^{-2.35}$), and a more "top-heavy" IMF (i.e. with a higher proportion of supersolar mass stars). The BC03 models offer the Chabrier ([2003]) IMF while the M05 models use the parametrization of Kroupa ([2001]). These latter IMFs are very similar and fit counts of low-mass stars in the Galaxy better than the Salpeter one (however, see Chapter 3). The minimum and maximum stellar masses considered are 0.1 $M_\odot$ and 100 $M_\odot$ respectively, which are consistent with the observed range of stellar masses. All three initial mass functions are shown in Fig. 2.1. Fig. 2.2 illustrates the sampling of a stellar library by the isochrone of a stellar population of 1 Gyr.

The integrated flux F_λ at a single wavelength of a single stellar population of a given age is then the sum of the fluxes at λ of all stars on the corresponding isochrone, weighted by the initial mass function (only for the main sequence, in the case of the M05 models; see above) :

$$F_\lambda = \int dm f_{\lambda,m} \phi(m) \qquad (2.1)$$

where $f_{\lambda,m}$ is the flux at λ of a star of mass m on the isochrone and can be normalized to a total mass of 1 $M_\odot$ in stars. This provides a convenient way of estimating the stellar masses of galaxies by comparing their fluxes with the model fluxes (see Section 2.2 and Chapter 3). It is important to remember that the model spectra obtained via this method do not reproduce, for example, the optical emission features in the spectra of star forming regions, as they are due to ionized gas. As absorption features can also be affected by

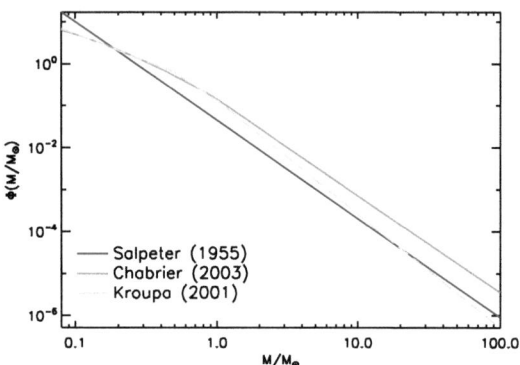

Figure 2.1: Initial mass functions of Salpeter ([1955], blue), Chabrier ([2003], red) and Kroupa ([2001], green), from $0.08 M_\odot$ to $100 M_\odot$.

emission infilling, these models are ill-suited to the study of star forming galaxies.

2.1.2 Composite stellar populations

While SSP models can reproduce the integrated spectra of star clusters accurately enough, a single burst of star formation is in general not a good descriptor of the star formation history of complex systems such as galaxies (e.g. Trager et al. [2000]). Throughout this work, we therefore adopted a more complex star formation history parametrized by a time-scale τ:

$$\psi(t) = \frac{t}{\tau^2} e^{-\frac{t}{\tau}} \qquad (2.2)$$

This delayed, exponentially declining star formation history is similar to the one proposed by Sandage ([1986]) and more realistic than a simpler exponentially declining star formation history (Gavazzi et al. [2002]). We will hereafter refer to models computed using this particular star formation history as "τ-models". These models are computed as the sum of a series of instantaneous bursts weighted by the star formation history (e.g. Tinsley [1980]). The spectral energy distribution at time T $F_\lambda(T)$ of a stellar population characterized by a star formation rate $\psi(t)$ is

$$F_\lambda(T) = \int_0^T dt \, \psi(T-t) f_\lambda(t) \qquad (2.3)$$

where $f_\lambda(t)$ is the spectrum of an SSP of age t. We call these models, obtained by combining several SSP templates using the star formation history described above, "composite stellar population" models.

2.1 Model SEDs and spectra

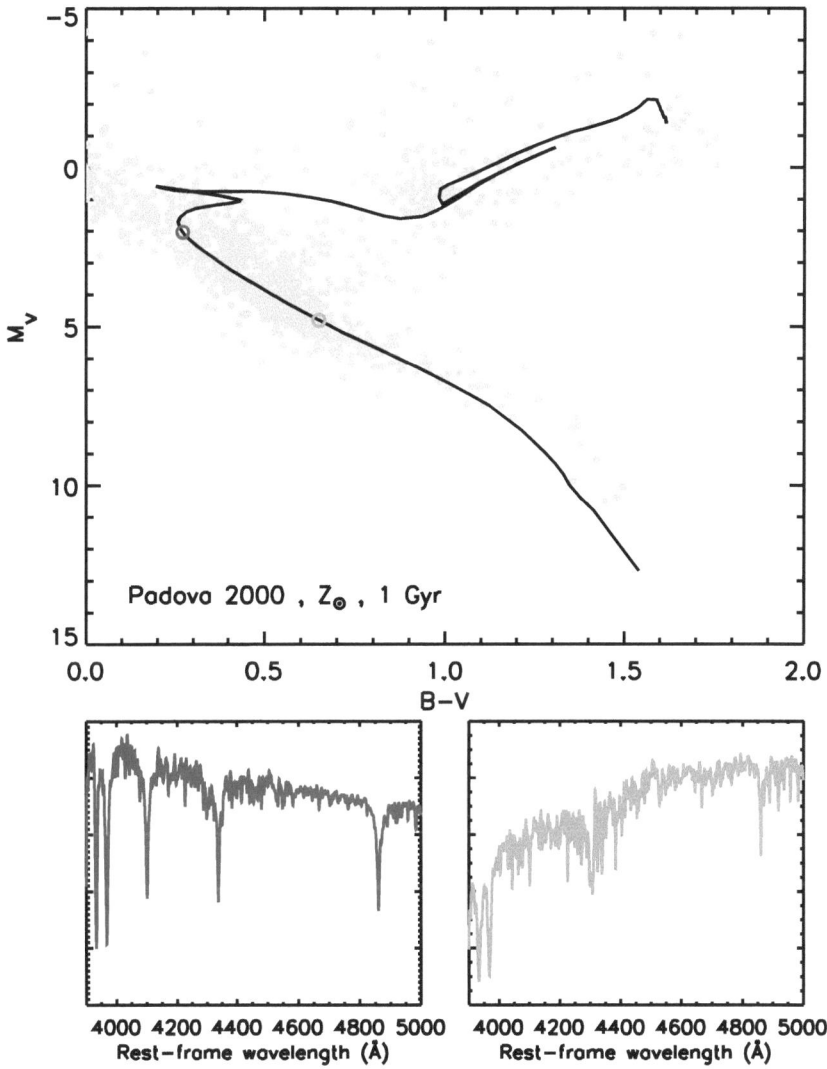

Figure 2.2: HR-diagram coverage of a stellar library (in this case, ELODIE; Prugniel & Soubiran [2001]) with a 1 Gyr old solar metallicity isochrone (Salasnich et al. [2000]) overlayed. Spectra of a F and G star are shown in blue and red respectively. The F star spectrum shows deep Balmer lines (see 2.2.1) but few metal ones while this is inverted in the G star spectrum.

In Fig. 2.3, we show the star formation history of a 4 Gyr old model population with a characteristic time-scale τ of 0.6 Gyr, with the spectra (around 4000 Å rest-frame) of some of the SSPs that make up the model (in color) as well as the composite spectrum within the same wavelength range (in black). The amplitude of the 4000 Å break (see below) is that of a stellar population of > 2 Gyr, the age of the bulk of the stars, but the hydrogen features (see below) are deeper than that of an SSP. This illustrates how this complex star formation history can be used to account for both an underlying old stellar population and a small amount of younger stars from a more recent burst.

The spectral energy distribution (SED), i.e. the flux in a given set of bandpasses, of a model is obtained by convolving the model spectrum with the response function of each band filter. The flux of the model spectrum in a given filter is then

$$F = \frac{\int_0^\infty d\lambda F_\lambda R(\lambda)}{\int_0^\infty d\lambda R(\lambda)} \quad (2.4)$$

where $R(\lambda)$ is the filter response function. This corresponds to the average flux in the wavelength range of the filter, weighted by the filter response.

2.2 Fitting the spectrophotometric data

2.2.1 Spectral features and bandpasses

In this work, we have used spectra taken in the i-band (from 6000 to 11000 Å) with FORS2 (Appenzeller & Rupprecht [1992]) on the ESO Very Large Telescope (VLT) and multiband, multi-observatory photometry in the optical and near infrared. In the redshift range around $z = 1$, that of the galaxies we studied, this corresponds to a rest-frame wavelength region around 4000 Å for the spectra and to the near-UV to near-IR for the photometry. In Fig. 2.4, we show the photometric bands and spectral features that characterize our data.

The region around 4000 Å contains several useful spectral features, the principal being the spectral break at 4000 Å for old stellar populations and the Balmer break at 3648 Å for young ones. The spectrum of a young ($\lesssim 1$ Gyr) stellar population is dominated by hot A and F type stars resulting in strong hydrogen absorption lines. As the stellar population ages, the main contribution to the flux shifts to cooler stars. The luminosity of the galaxy decreases, as does the depth of the Balmer lines, and the drop at the end of the Balmer series is replaced with a spectral break at 4000 Å due to blanketing by metal lines. The strength of the 4000 Å break increases with age and metal content and is, for a fixed metallicity, a measure of age (e.g. Poggianti & Barbaro [1997], Kauffmann et al. [2003]) and the telltale indicator of an old population. The presence of deep Balmer lines, on the other hand, is the signature of a young stellar population (e.g. Couch & Sharples [1987], Poggianti et al. [1999]). Because of the aforementioned veil of metal lines, the true continuum can not be measured (at the resolution of our FORS2 spectra anyway) and the apparent

2.2 Fitting the spectrophotometric data

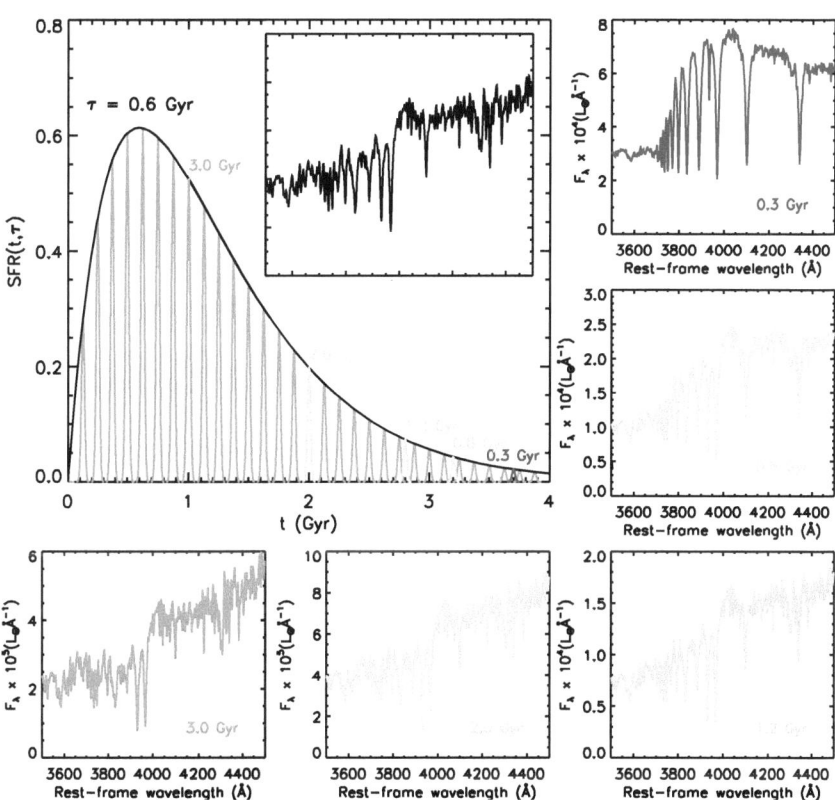

Figure 2.3: Star formation history of a 4 Gyr old τ-model with $\tau = 0.6$ Gyr. The inserts show some of the SSP templates that compose the model, in color, and the resulting composite spectrum, in black. The spectra are shown in the wavelength range around 4000 Å. The young SSP spectra have deep Balmer lines but few metal ones, while in the old SSP spectra the metal features dominate. The composite spectrum shows a mix of moderately deep Balmer lines and prominent metal features.

depth of the Balmer features varies with the metal content of the stellar population. In general, all high-order Balmer lines from Hδ down to the Balmer limit are present in our spectra. Another prominent feature in this wavelength region that can be used to estimate the age of the galaxy's stellar population is the Ca II H & K doublet at 3968 Å and 3934 Å respectively. As the Ca H line is blended with the high-order Balmer line Hϵ, the ratio between Ca H and Ca K is a measure of the presence of late-B to early-F stars (Rose [1985]).

For all the samples used in this work, we had broadband photometry in the i, z, J and K bands (~ 0.77, ~ 0.9, ~ 1.2 and ~ 2.2 μm respectively). In all cases two filters straddled the 4000 Å break, so that the resulting color was as a rough age indicator, for a given redshift and metallicity, and could be used to separate early-type galaxies from star forming ones. As the infrared emission is dominated by long-lived stars and is therefore insensitive to ongoing star formation, the flux in the K band (or higher wavelength bandpasses) provides a good constraint on the galaxies' stellar mass (e.g. Kauffmann & Charlot [1998], Rettura et al. [2006]). On the other hand, massive short-lived stars emit most of their light in the rest-frame UV, making the B-band (or a shorter wavelength band), when available, a good tracer of residual star formation that can be used to constrain the star formation history of an early-type galaxy. We see that the photometric and spectroscopic data are complementary. The SED covers a wide range of wavelengths and provides information on mass and current star formation history while the spectrum, although on a much more limited wavelength range, allows one to determine the age of the galaxy's stellar population with greater precision. The combination of photometry and spectroscopy therefore puts stronger constraints on the star formation history than either alone.

2.2.2 Spectrophotometric fitting method

In the rest-frame wavelength range investigated here, the resolution of the BC03 templates is 3 Å and that of the M05 templates ~ 15 Å. When the model spectra are redshifted at $z \sim 1$, the resolution of the M05 templates becomes lower than that of the FORS2 spectra we used (~ 12 Å). The resolution of our observed spectra would thus need to be degraded in order for them to be compared to models computed from M05 SSPs. For this reason, we decided to compute our composite stellar population models from BC03 SSPs instead. We did however use M05 models to compare stellar mass estimates (see Chapter 3). From 2.2.1, it becomes clear that the effect of age on the SED and spectrum (a deeper 4000 Å break, shallower Balmer features) can be reproduced assuming a different metal content. This results in an anticorrelation between age and metallicity, the well-known **"age-metallicity degeneracy"**. This is illustrated in Fig. 2.5, where we plot the SED and spectra of a solar and suprasolar metallicity model. The model SED and spectra are very similar, despite an age difference of several Gyr between the two models. For this reason, when comparing composite stellar population models to the observed data, we assumed a constant metallicity of the models. In addition to solar metallicity templates, the BC03 models offer the choice of several subsolar metallicities and one suprasolar (Z=2.5Z$_\odot$) metallicity. As the metal content of early-type galaxies appears to be slightly higher than

2.2 Fitting the spectrophotometric data

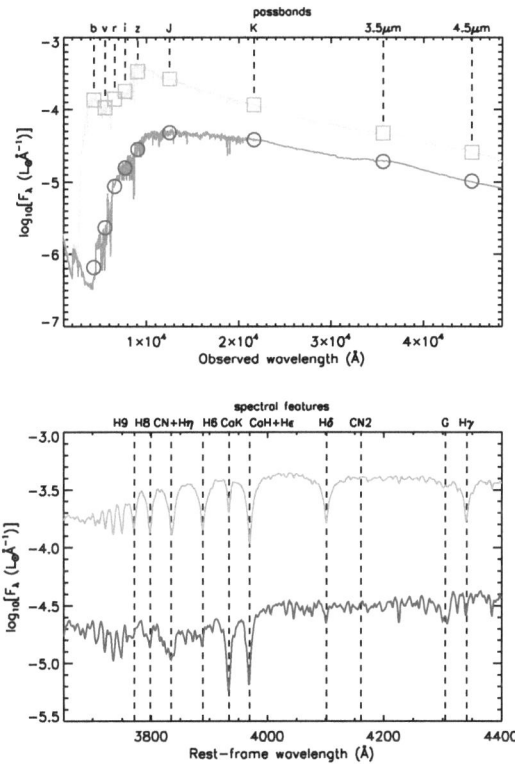

Figure 2.4: Relevant passbands and spectral features in the wavelength ranges used in this work. Top: model SED and spectrum of a 4 Gyr (blue circles, dark grey) and 0.5 Gyr (red squares, light grey) single stellar population model redshifted to $z \sim 1$. Bottom: spectra of the same two models, in the rest-frame wavelength range of 3600 to 4400 Å. The effective wavelength of each bandpass and the spectral features are indicated by dashed lines.

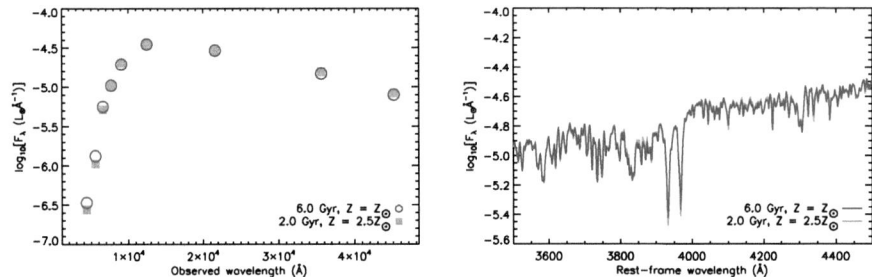

Figure 2.5: Age-metallicity degeneracy: comparison of the SED and spectra of two single stellar population BC03 templates at solar (blue) and 2.5 solar (red) metallicity and with ages of 2 and 6 Gyr respectively. The spectrophotometry of the two models is remarkably similar although the suprasolar metallicity model is 4 Gyr younger than the solar metallicity one.

solar to twice the solar value (e.g. Gallazzi et al. [2006], Jimenez et al. [2008]) and because we expect the stellar library to be more complete at solar metallicity, we always considered solar metallicity models first.

The resolution of the BC03 templates redshifted to $z \sim 1$ (6 Å) is still twice as high as the resolution of the FORS2 spectra used in this work. We therefore downgraded the resolution of our model spectra to that of the observed ones using a Gaussian broadening function. The broadened model spectrum F'_λ is then given by

$$F'_\lambda(\lambda) = \frac{1}{\sigma\sqrt{2\pi}} \int_0^\infty d\lambda' F_\lambda(\lambda') e^{-\frac{(\lambda-\lambda')^2}{2\sigma^2}} \quad (2.5)$$

where F_λ is the original spectrum and $\sigma = \sqrt{\Gamma^2_{FORS2} - \Gamma^2_{STELIB}}/2.3548$, with Γ_{FORS2} and Γ_{STELIB} being the FWHM resolution of the observed and model spectra respectively. Also, to account for the broadening of lines due to the distribution of stellar velocities in a galaxy, we applied a Gaussian velocity dispersion to our composite stellar population spectra:

$$F'_\lambda(\lambda) = \frac{1}{\sigma_v\sqrt{2\pi}} \int_{-\infty}^\infty dv F_\lambda\left(\lambda(1+\frac{v}{c})^{-1}\right) e^{-\frac{v^2}{2\sigma_v^2}} \quad (2.6)$$

where c is the speed of light and σ_v the stellar velocity dispersion, which we let vary between 0 and 400 km/s. Finally, each model spectrum was interpolated at the wavelenghts of the observed spectrum it was compared with.

We derived stellar population parameters for our sample galaxies by comparing the grid of composite stellar population models described above with the observed SEDs and spectra.

2.2 Fitting the spectrophotometric data

This was done by minimizing a chi-square (χ^2) estimator, defined for the photometry as

$$\chi^2(T, \tau, M) = \sum_i \frac{(F_{i,o} - M \times F_i(T, \tau))^2}{\sigma_i^2} \qquad (2.7)$$

where $F_{i,o}$ and σ_i are the observed flux and flux error in the i-th band respectively, $F_i(T, \tau)$ is the flux in the i-th band of the $\{T, \tau\}$ model spectrum, per solar mass, and M is the stellar mass of the model. For the fit to the observed spectrum, the χ^2 is

$$\chi^2(T, \tau, \sigma_v) = \sum_\lambda \frac{(F_{\lambda,o} - F_\lambda(T, \tau, \sigma_v))^2}{\sigma_\lambda^2} \qquad (2.8)$$

where $F_{\lambda,o}$ and $F_\lambda(T, \tau, \sigma_v)$ are the flux value at λ of the observed and model spectrum respectively and σ_λ is the flux error at λ. This latter value was derived from the signal-to-noise ratio (S/N) of the observed spectrum. As absorption spectra of old populations, such as those of early-type galaxies, have very few true continuum regions, we estimated the S/N from the residuals of fitting the H_δ absorption feature with the combination of a Gaussian profile and a first-degree polynomial. Since we expect the true star formation history of a galaxy to be more complex than a simple delayed exponential (e.g. Marri & White [2003], De Lucia et al. [2006]), and because the spectra of galaxies at $z \sim 1$ are often very noisy, the best fit models are likely to be not sufficient to properly describe the actual star formation history of the studied galaxies. And as the photometry and spectroscopy are likely to be affected by different systematic uncertainties, the best fits to the observed SED and spectrum might not coincide. Therefore, in order to cover most of the star formation histories associated with the observed data for a given set of models, we considered all models within both the 99.7% (hereafter, "3σ") confidence regions, in the space of model parameters, of the fit to the SED and spectrum. We will hereafter refer to these models as "best fitting" models.

For a χ^2 statistic, the 3σ confidence region is defined as $\chi_\alpha^2 = \chi_0^2 + \Delta(\nu, \alpha)$ (Avni [1976]), where χ_0^2 is the minimum χ^2 value from the fit and $\Delta(\nu, \alpha)$ is such that $P(\chi^2 \leq \Delta(\nu)) = \alpha$. Here $\alpha = 0.997$ and ν is the number of degrees of freedom, i.e. the number of independent variables minus the number of free parameters, the latter being T, τ and M (for the fit to the SED) or σ_v (for the fit to the spectrum). In the case of the SED, the variables are the fluxes in each bandpass. For the spectrum however, the data points can not be considered as independent variables since the wavelength sampling is in this case finer than the actual resolution. As our spectra show little to no continuum in the age and wavelength ranges considered, we assume the independent variables to be the spectral features fitted by our models (some of which are shown in Fig. 2.4). To test this hypothesis, we performed a set of Monte Carlo simulations on each τ-model in a grid with T from 0.2 to 5 Gyr and τ from 0 to 2 Gyr. We perturbed the SED and spectrum of each model a thousand times, assuming bandpasses and random normal errors consistent with the data, and fitted each separately using the same grid of models. We then compared the results of the simulations with those of a fit using the same parameter grid but to unperturbed

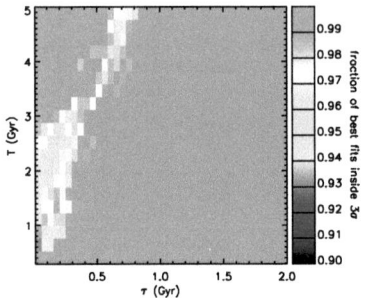 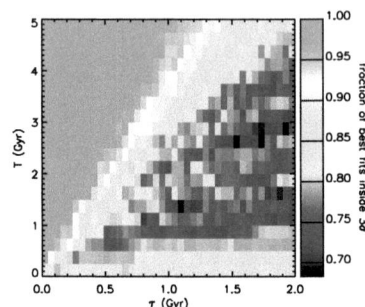

Figure 2.6: Fraction of best fit results from the Monte-Carlo simulations within the 3σ confidence region of the fit to the spectrum (left) and SED (right) of each model in the grid, for values of T from 0.1 to 5 Gyr and τ from 0.01 to 2 Gyr, assuming wavelength ranges and errors consistent with the observed data (see following Chapters).

models. In Fig. 2.6, we show for each model in the grid the fraction of best fit models (i.e. with $\chi^2 = \chi_0^2$) from the simulation that are within the 3σ confidence region of the fit to the unperturbed model. We found that while the confidence regions of both fits were systematically different, even if slightly, for old $(T - t_{fin} \gtrsim 2.5$ Gyr, see Section 2.3) models the correction to $\Delta(\nu)$ is small enough as to not change the overall distribution of models within the confidence region, thus validating our assumption. For models with younger ages, the confidence regions of the fit are underestimated and would need to be corrected. We note that, in this work, the best fit τ-models to the spectrophotometric data of our early-type galaxies almost always fell within the "old models" region of parameter space defined above (the mean star formation history parameters of the models within 3σ can however fall outside this region). When a more complex star formation history was used (see Chapter 4) or a different wavelength range (see Chapter 6) we carried out the same test, but on the best fit models to the data only.

Next we estimated the constraining power of our spectrophotometric data. Because the quality of the photometry used in this work was relatively constant, we focused on the spectroscopy. We performed the fitting procedure on each of the τ-models in a grid with T varying from 0 to 5 Gyr and τ from 0 to 2 Gyr. We used the same set of filters and wavelength range as for the observed data and assumed photometric errors comparable to those of the data, but let the S/N of the model spectra vary. In Fig. 2.7, we plot as a function of T and τ the maximum S/N for which the confidence region of the fit to the simulated spectrum completely overlaps the confidence region of the fit to the simulated SED, i.e. the maximum S/N for which the spectroscopy does not add further constraints on the star formation histories than those given by the photometry. We found that on average a S/N of 6.5, and not lower than 4, is needed in order for the spectroscopy to be

2.3 Characterizing the star formation history

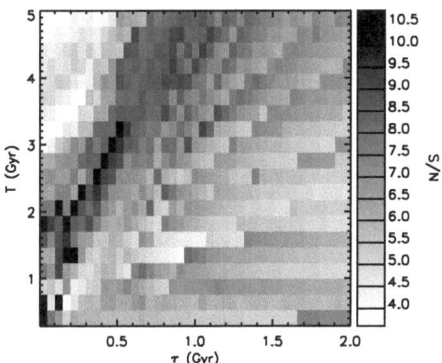

Figure 2.7: Distribution, as a function of τ and T, of the minimum signal to noise ratio of the spectrum required to add further constraints on the star formation histories to those given by the SED, assuming standard photometric errors.

useful. As many of the individual galaxies studied in this work have a lower S/N, this shows the need to group the galaxies together and fit their averaged (stacked) spectrum instead.

2.3 Characterizing the star formation history

To characterize the star formation history of a given model, we used two different age like estimators. The first is the star formation weighted age, defined as

$$t_{SFR}(T,\tau) = \frac{\int_0^T dt(T-t)\psi(t,\tau)}{\int_0^T dt\psi(t,\tau)} \qquad (2.9)$$

where $\psi(t,\tau)$ is the star formation rate of the model, as a function of the time since the onset of star formation, as defined in Eq. 2.2. This definition takes into account the effective fraction of stellar mass contributed by each single stellar population making up the model and stellar populations contributing only a negligible fraction of the final stellar mass (i.e. the stellar mass at T) do not affect the star formation weighted age significantly. For an instantaneous burst of star formation at $t = 0$, such as in an SSP, $t_{SFR} = T$ while for a constant star formation rate, $t_{SFR} = T/2$.

We also used a second estimator, which we call the final formation time t_{fin}, defined as the time after the onset of star formation at which the stellar mass is a (large) fraction of the final stellar mass $M^*(T)$. Here we chose

$$M^*(t_{fin}) = 0.99 \times M^*(T) \qquad (2.10)$$

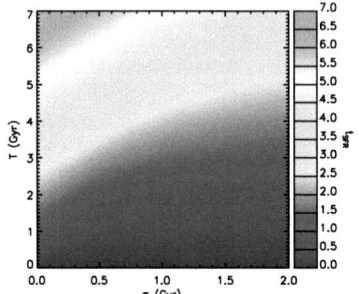 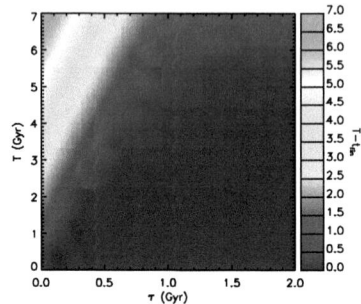

Figure 2.8: Distribution of t_{SFR} (left) and $T - t_{fin}$ (right) of τ-models, for values of T from 0.1 to 7 Gyr and τ from 0.01 to 2 Gyr. t_{SFR} varies smoothly across the grid while $T - t_{fin}$ is much more sensitive to residual star formation in old models (i.e. small values of τ compared to T).

Unlike the star formation weighted age t_{SFR}, the final formation time t_{fin} is sensitive to residual star formation. So while t_{SFR} measures the age of the bulk of the stars in a galaxy, t_{fin} traces the last stages of stellar mass assembly and is therefore useful to distinguish between two otherwise old stellar populations that have stopped star formation at different times. For a model that fits the observed SED or spectrum of a galaxy, $T - t_{fin}$ is the look-back time from the epoch of the galaxy to the last episode of star formation and is independent of the time at which the star formation of the model started. In Fig. 2.8, we plot the star formation weighted age t_{SFR} and the look-back time to the final formation $T - t_{fin}$ of τ-models as a function of T and τ.

To check that the spectrophotometric fit is not biased when fitting actual (i.e. noisy) data, we compared the mean t_{SFR} and $T - t_{fin}$ of the fit to the unperturbed models, $\overline{t_{SFR}}$ and $\overline{T - t_{fin}}$, with the averages of theses values from the fits to the perturbed models, $\langle \overline{t_{SFR}} \rangle$ and $\langle \overline{T - t_{fin}} \rangle$ respectively. We found no significant bias when using the model grid to fit noisy data with respect to model SEDs and spectra (see Fig. 2.9). We also compared the results of the fit done with BC03 models to the fit done using M05 models. In Fig. 2.10, we plot the difference between the mean star formation weighted ages and final formation look-back times of the best fitting BC03 models to the BC03 grid and those of the best fitting M05 models to the M05 grid. We found that the difference was smaller than 0.1 Gyr, except in the parameter range where TP-AGB stars dominate. This shows that, as long as the stellar population considered is older than $\sim$2 Gyr, the spectrophotometric fit does not show a significant bias when using a set of models over the other.

2.3 Characterizing the star formation history

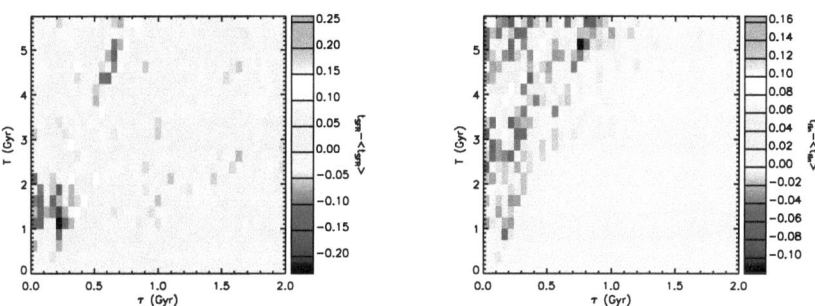

Figure 2.9: Difference between the mean t_{SFR} (left) and $T - t_{fin}$ (right) of best fitting models to the input model and the average of the same parameter from Monte Carlo simulations.

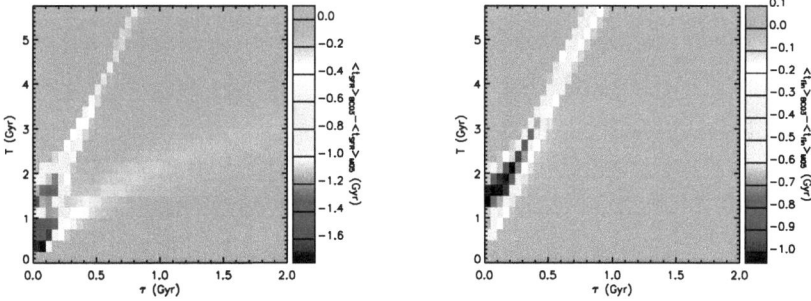

Figure 2.10: Difference between the mean t_{SFR} (left) and $T - t_{fin}$ (right), of best fitting models to the input model, using BC03 and M05 templates. Note how the two sets of models yield different results in the age range where the contribution of TP-AGB stars is maximal.

2.3.1 Effects of metallicity

As the metallicity of early-type galaxies may vary, in particular with mass (e.g. Bernardi et al. [2005], Thomas et al. [2005]), Sánchez-Blázquez, et al. [2006]), keeping the metallicity fixed when fitting models to the observed data can induce systematic errors in the stellar population parameters. This is especially relevant when two galaxy samples are being compared, as a metallicity difference between the galaxy populations can produce an apparent (and spurious) difference in stellar population properties. To quantify this bias, we carried out another test, in which we fitted a BC03 model at solar metallicity to a series of subsolar and suprasolar metallicity BC03 models computed from the same parameter grid and assuming photometric and spectroscopic errors consistent with the observed data. The nonsolar metallicity models were obtained by interpolating a set of three models with $Z = 0.4Z_\odot$, $Z_\odot$ and $2.5Z_\odot$ respectively at different metallicities, from 0.5 to $2Z_\odot$. In Fig. 2.11 and 2.12 we show, as a function of the parameters T and τ, the difference between the mean star formation weighted age, and final formation time respectively, of the best fitting solar metallicity models to the solar metallicity input and that of the best fitting solar metallicity models to the nonsolar metallicity ones, which we note Δt_{SFR} and Δt_{fin}. Unsurprisingly, the bias is significant for old models, which have prominent metal features, while it is negligible for young models with spectra dominated by Balmer absorption. For models with $T - t_{fin} \gtrsim 1.5$ Gyr, we found that Δt_{SFR} varies as $0.7(\pm 0.1)Z/Z_\odot$ and Δt_{fin} as $0.85(\pm 0.1)Z/Z_\odot$. Interestingly, at suprasolar metallicities the difference of t_{SFR} and t_{fin} is maximal for models whose age is ~ 2 Gyr and decreases for older models. As this effect disappears when performing the same test on M05 models, it is likely not intrinsic to the fitting procedure (i.e. caused by the boundaries of the parameter grid, for example) but due to a particularity of the BC03 templates. The age at which the difference is maximal suggests that this an effect of the particular treatment of post-main sequence stars in the BC03 model.

2.4 Summary

In this Chapter, we have described a method to estimate star formation histories consistent with observed photometric and spectroscopic data using a set of composite stellar population synthesis models. By comparing this grid of models to both the SED and spectrum of galaxies, we obtain a subset of models that reproduce the broadband colors as well as the spectral features of these galaxies. We extensively tested this method of spectrophotometric fitting:

- we checked that the assumptions made in the fitting procedure were essentially correct for the range of stellar population parameters considered in this work (i.e. those of old passive populations).

- we discussed two different sets of single stellar population templates, those of Bruzual & Charlot ([2003]) and those of Maraston ([2005]) and their respective strengths and

2.4 Summary

weaknesses. For the study of observed spectra, we decided to use the Bruzual & Charlot ([2003]) models, on account of their superior spectral resolution.

- we investigated the possible biases resulting from our fitting approach and the choice of models and found these to be negligible, except in a narrow age range were stars on the asymptotic giant branch dominate the integrated light of the model stellar population.

- finally, we quantified the effect of metallicity on the stellar population parameters, which was found to be significant when old stellar populations are considered.

For these tests, and in order to characterize the star formation histories of galaxies, we defined two complementary age estimators, namely the star formation weighted age and the final formation time. The former measures the age of the bulk of the stars of the galaxy while the latter traces the last significant episode of star formation. From these checks, we concluded that the combined analysis of observed spectrophotometric data by the mean of composite stellar population models is suited to study the star formation histories of now passively evolving galaxies.

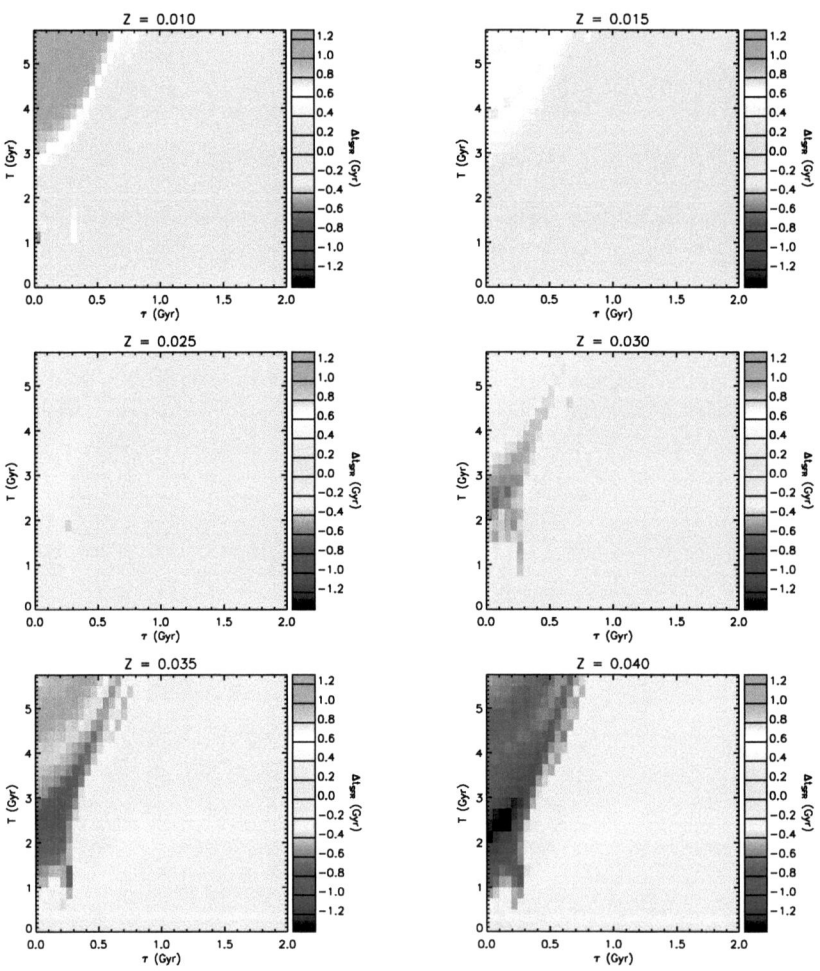

Figure 2.11: Metallicity bias for t_{SFR}: difference in mean t_{SFR} between the best fitting solar metallicity models to solar and nonsolar metallicity inputs. Old models (top left corner of the grid) at subsolar (respectively suprasolar) metallicity appear younger (respectively older) than solar metallicity models while the fit to younger models is unaffected by metallicity.

2.4 Summary

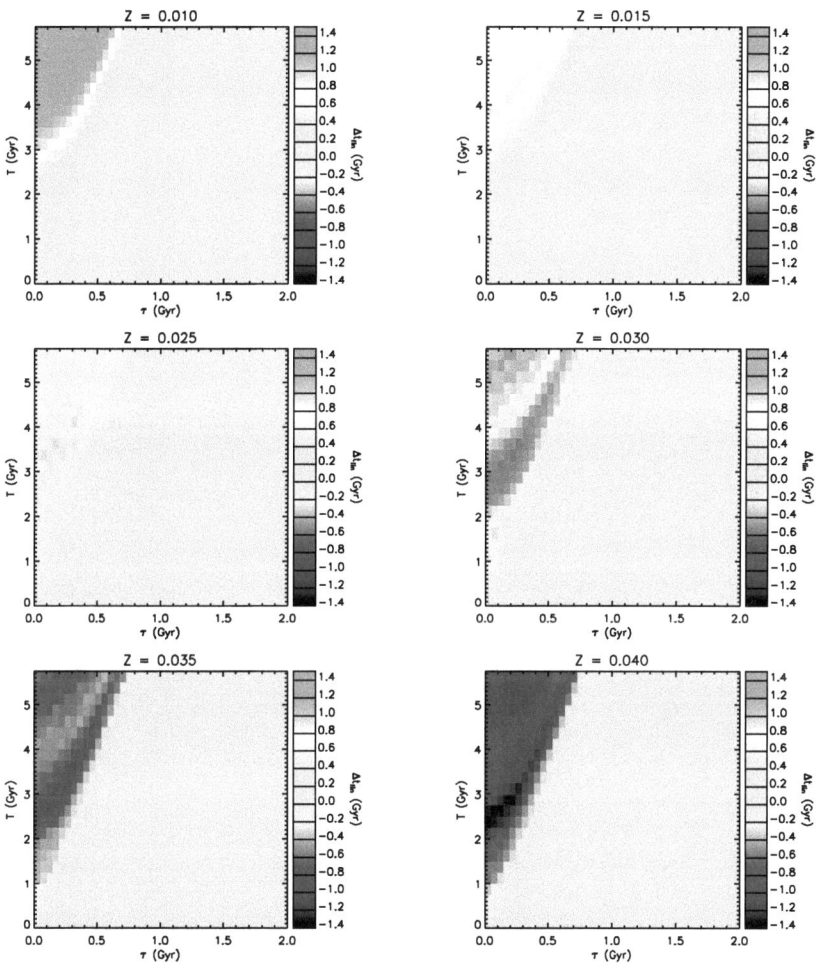

Figure 2.12: Metallicity bias for $T - t_{fin}$: difference in mean $T - t_{fin}$ between the best fitting solar metallicity models to solar and nonsolar metallicity inputs. Old models (top left corner of the grid) at subsolar (respectively suprasolar) metallicity appear younger (respectively older) than solar metallicity models while the fit to younger models is unaffected by metalliticy.

Chapter 3

Robustness of stellar mass estimates

An estimate of the mass contained in galaxies in the form of stars is interesting for several reasons. By the comparison of photometric mass estimates, obtained for example by spectral energy distribution (SED) fitting methods (see Chapter 2), with dynamical or lensing measurements, it is possible to study the radial distribution of dark matter in galaxies (e.g. Ferreras et al. [2005], [2008], Napolitano et al. [2005]), to investigate the relationship between visible and dark matter (e.q. Lintott et al. [2006], Rettura et al. [2006]) and to test hierarchical structure formation models (e.g. Nagamine et al. [2004], De Lucia et al. [2006]). Furthermore, the evolution of galaxies is known to depend on their stellar mass, with massive galaxies appearing older than their less massive counterparts (e.g. Cimatti et al. [2006], Holden et al. [2007], Pozzetti et al. [2007]). Accurate estimates of the stellar content in galaxies can therefore be used to disentangle effects due to mass from those due to environment when comparing different galaxy populations. Interestingly, Treu & Koopmans ([2004]) have proved that the fraction of mass in the form of stars in elliptical lens galaxies can also be estimated with a joint lensing and dynamical analysis.

Although the stellar mass is often measured using one of these techniques, only a few studies have been performed to check the reliability of each method (e.q. Drory et al. [2004], Rettura et al. [2006], van der Wel et al. [2006]). Further investigations are therefore important to probe the consistency of these different techniques.

This Chapter is organized as follows. In Section 3.1, we describe two samples of elliptical lens galaxies from the Sloan Lens ACS Survey (SLACS). In Section 3.2, we describe stellar mass estimates based on lensing and dynamical methods. In section 3.3, we present stellar mass estimates obtained from SED modeling, for a sample of 15 elliptical galaxies from SLACS. In Section 3.4, we discuss the consistency of the results obtained by those two diagnostics. In Section 3.5, we present an analysis of the amount and distribution of dark matter in the elliptical galaxies of the SLACS and SLACS II samples, by combining the photometric stellar mass measurements with lensing measurements of the total mass.

3.1 The SLACS sample

The Sloan Lens ACS Survey (or SLACS) is a spectroscopic and imaging survey of new early-type strong gravitational lenses. Here we summarize the selection process, described in Bolton et al. ([2004], [2006]). Lens candidates were spectroscopically selected from the luminous red galaxy (LRG, Eisenstein et al. [2001]) sample of the Sloan Digital Sky Survey (SDSS, York et al. [2000]) for having multiple nebular emission lines at a significantly higher redshift than the target's. This was done by fitting a Gaussian line profile to the residuals of the continuum-subtracted spectrum. From the SDSS 3" aperture stellar velocity dispersion and the "foreground" and "background" redshifts, Bolton et al. ([2006]) estimated the strong lensing probability of the candidates using a singular isothermal sphere model (see Section 3.2). The most promising systems were then observed in the F435W (b_{435}) and F814W (i_{814}) bands with the Advanced Camera for Surveys (ACS, Ford et al. [1998])on the Hubble Space Telescope (HST) to confirm the lensing hypothesis. The candidate systems whose residual image, after subtracting a fitted model to the target galaxy luminosity profile, could be interpreted in terms of a simple lens model were classified as strong gravitational lenses. Finally, a lensing analysis was performed on the i_{814} images. Treu et al. ([2006]) and Koopmans et al. ([2006]) first performed a detailed lensing and dynamical analysis on a subsample of 15 early-type lens galaxies. In a second time, Gavazzi et al. ([2007]) performed a weak lensing analysis of 12 new lenses. These subsamples will be hereafter called SLACS and SLACS II respectively. Fig. 3.1 shows some of the lensing systems in the SLACS sample. Table 3.1 and 3.2 list the lens galaxies in the SLACS and SLACS II samples, with some of their relevant properties.

An analysis of the photometric and structural parameters of these lenses revealed that the lens galaxies are a representative sample of the SDSS sample with respect to color and ellipticity but that they are significantly biased towards higher surface brightness, at a fixed redshift and velocity dispersion.

3.2 Stellar mass estimates

We derived the stellar masses of the lens galaxies in the SLACS and SLACS II samples by comparing their SDSS photometry to composite stellar population models. We then compared our photometric stellar mass estimates for the SLACS sample to the stellar masses estimated by Koopmans et al. ([2006]) using a combination of strong lensing modeling and dynamical analysis. In this Section, we briefly describe their approach and the mass estimates from strong lensing and stellar dynamics. For a complete treatment of strong gravitational lensing and galactic dynamics respectively, we refer to Schneider et al. ([1992]) and Binney & Tremaine ([1987]).

3.2 Stellar mass estimates

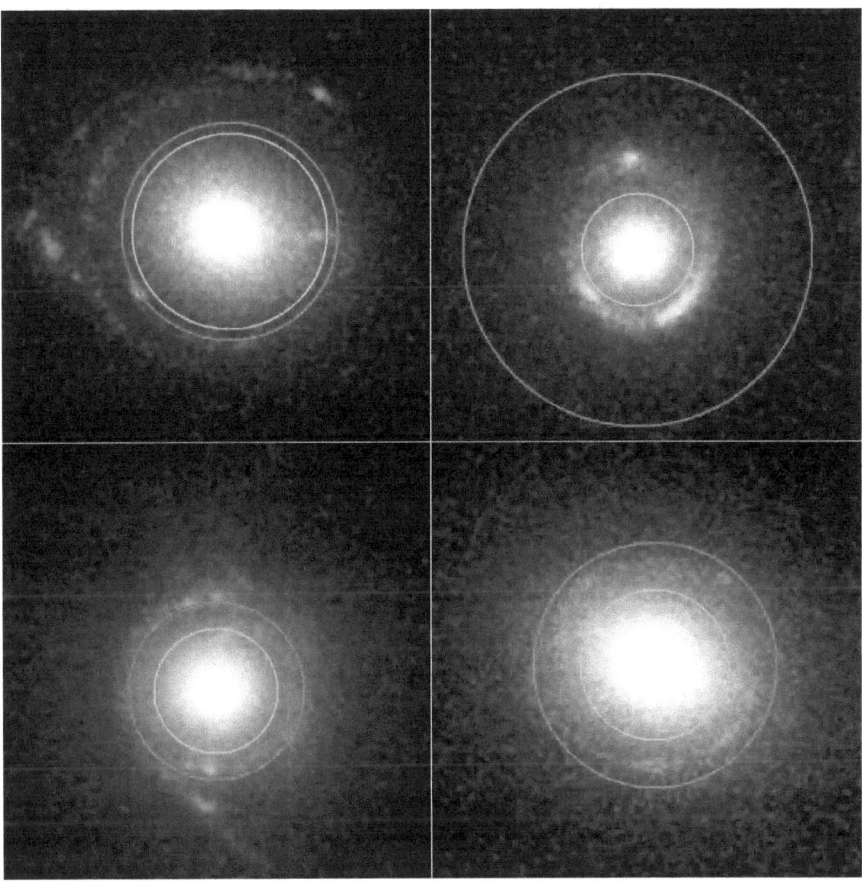

Figure 3.1: Multiband (b_{435} and i_{814}) images of four lens galaxies used in this study (Bolton et al. ([2006])). Each image is 8" wide (Credit: NASA, ESA and the SLACS survey team). The green circles represent the Einstein radius of each lens galaxy and the red circle its effective radius. The variation, among the elliptical lens galaxies, of the Einstein radius with respect to the effective radius allowed us to study the average dark matter distribution within these galaxies (see Section 3.5).

SDSS ID	z_{lens}	z_{source}	R_e (")	R_{Ein} (")
J0037-0942	0.1955	0.6322	2.38	1.47
J0216-0813	0.3317	0.5235	2.79	1.15
J0737+3216	0.3223	0.5812	3.26	1.03
J0912+0029	0.1642	0.3240	5.50	1.61
J0956+5100	0.2405	0.4700	2.60	1.32
J0959+0410	0.1260	0.5349	1.99	1.00
J1250+0523	0.2318	0.7950	1.64	1.15
J1330-0148	0.0808	0.7115	1.23	0.85
J1402+6321	0.2046	0.4814	2.29	1.39
J1420+6019	0.0629	0.5352	2.49	1.04
J1627+0053	0.2076	0.5241	2.47	1.21
J1630+4520	0.2479	0.7933	2.01	1.81
J2300+0022	0.2285	0.4635	1.80	1.25
J2303+1422	0.1553	0.5170	3.73	1.64
J2321-0939	0.0819	0.5324	4.47	1.58

Table 3.1: Relevant properties of the SLACS sample: redshifts of the lens galaxy, redshift of the lensed galaxy, effective radius and Einstein radius (Koopmans et al. [2006], Gavazzi et al. [2007]).

SDSS ID	z_{lens}	z_{source}	R_e (")	R_{Ein} (")
J0029-0055	0.227	0.931	1.48	0.82
J0157-0056	0.513	0.924	0.93	0.72
J0252+0039	0.280	0.928	1.69	0.98
J0330-0020	0.351	1.107	1.17	1.06
J0728+3835	0.206	0.688	1.33	1.25
J0808+4706	0.220	1.025	1.65	1.23
J0903+4116	0.430	1.065	1.28	1.13
J1023+4230	0.191	0.696	1.40	1.30
J1103+5322	0.158	0.735	3.22	0.84
J1205+4910	0.215	0.481	1.92	1.04
J2238-0754	0.137	0.713	2.33	1.20
J2341+0000	0.186	0.807	3.20	1.28

Table 3.2: Relevant properties of the SLACS II sample: redshifts of the lens galaxy, redshift of the lensed galaxy, effective radius and Einstein radius (Gavazzi et al [2007]).

3.2 Stellar mass estimates

3.2.1 Lensing mass

According to the theory of General Relativity (GR), light passing at distance ξ from a point mass M is deflected by an angle

$$\hat{\alpha} = \frac{4GM}{c^2 \xi} \qquad (3.1)$$

in the weak field limit, i.e. as long as the gravitational field and velocity of the deflecting mass are small compared to c. The true angular position of the light source with respect to the lensing mass, $\mathbf{y}$, is related to its observed position $\mathbf{x}$ by the ray-tracing equation

$$\mathbf{y} = \mathbf{x} - \frac{D_{ls}}{D_{os}} \hat{\boldsymbol{\alpha}}(D_{ol}\mathbf{x}) = \mathbf{x} - \boldsymbol{\alpha}(\mathbf{x}) \qquad (3.2)$$

where D_{ls} is the distance of the lens to the source, D_{os} the distance of the observer to the source and D_{ol} the distance of the observer to the lensing mass. Note that, while the shape of the source is not conserved by lensing, its surface brightness is. In the limit of weak fields, the GR equations can be linearized and the deflection angle of a lens made of a distribution of point masses is the sum of the deflection angles of the individual mass components. In integral form,

$$\boldsymbol{\alpha}(\mathbf{x}) = \frac{1}{\pi} \int d^2x' \kappa(\mathbf{x}') \frac{\mathbf{x} - \mathbf{x}'}{\|\mathbf{x} - \mathbf{x}'\|^2} \qquad (3.3)$$

where $\kappa(\mathbf{x})$ is the dimensionless surface mass density defined as

$$\kappa(\mathbf{x}) = \frac{\Sigma(D_{ol}\mathbf{x})}{\Sigma_{cr}} \qquad (3.4)$$

Σ is the surface mass density and Σ_{cr} the critical surface mass density defined as

$$\Sigma_{cr} = \frac{c^2}{4\pi G} \frac{D_{os}}{D_{ol} D_{ls}} \qquad (3.5)$$

If $\Sigma \geq \Sigma_{cr}$, then source positions $\mathbf{y}$ exist such that a source at $\mathbf{y}$ has multiple images. For a point source on the same line of sight as the lensing mass, $\mathbf{y} = 0$ and the image of the source is a circle of radius R_{Ein} (or in angular terms, θ_{Ein}, where $R_{Ein} = D_{ol}\theta_{Ein}$), called the Einstein radius (respectively angle), such that the mean surface mass density Σ enclosed within R_{Ein} is equal to the critical mass density Σ_{cr}. The projected mass of the lens enclosed within this circle is then

$$M_{len}(\leq R_{Ein}) = \Sigma_{cr} \pi R_{Ein}^2 \qquad (3.6)$$

A simple model that is commonly used is the singular isothermal sphere (hereafter, SIS), defined by the three-dimensional density distribution

$$\rho(r) = \frac{\sigma_v^2}{2\pi G r^2} \qquad (3.7)$$

and characterized by a one-dimensional velocity dispersion σ_v. In the case of a SIS, the Einstein radius is given by

$$R_{Ein} = 4\pi \left(\frac{\sigma_v}{c}\right)^2 \frac{D_{ol}D_{ls}}{D_{os}} \quad (3.8)$$

which corresponds to an Einstein angle of

$$\theta_{Ein} \approx 2.6 \left(\frac{\sigma_v}{300}\right)^2 \frac{D_{ls}}{D_{os}} \text{arcsec} \quad (3.9)$$

for σ_v in km/s. As the SLACS lenses are all at low to intermediate redshift and the source cannot be much further than $z \sim 1$ (or it would be too faint), the quantity D_{ls}/D_{os} is of the order of unity. For a lens at $z = 0.1$, a source at $z = 1$ and a velocity dispersion of 200 km/s, for example, $\theta_{Ein} \sim 1$". A generalization of the SIS, called the singular isothermal ellipsoid (SIE) was used by the SLACS team to successfully model the lenses in the SLACS sample (Bolton et al. [2006], Treu et al. [2006]). Treu et al. ([2006]) have found that the velocity dispersion σ_v of the best fit SIE lensing models to the SLACS lenses approximates very well the central velocity dispersion σ_0 of the lensing galaxies.

3.2.2 Dynamical mass

The stellar component of a galaxy is best described as a collisionless system where the stars move under the influence of the mean gravitational potential $\Phi(\mathbf{x}, \mathbf{v}, t)$ of the galaxy. Because there are no collisions, the density of stars $f(\mathbf{x}, \mathbf{v}, t)$ satisfies the continuity equation

$$\frac{\partial f}{\partial t} + \sum_{i=1}^{6} \frac{\partial f \dot{w}_i}{\partial w_i} = 0 \quad (3.10)$$

where $\mathbf{w} = (\mathbf{x}, \mathbf{v})$ and $\dot{\mathbf{w}} = (\mathbf{v}, -\nabla\Phi)$. Since v_i and x_i are independent coordinates of phase space and $\nabla\Phi$ does not depend on the velocities, Eq. 3.10 can be simplified into the collisionless Boltzmann equation

$$\frac{\partial f}{\partial t} + \sum_{i=1}^{3} \left(v_i \frac{\partial f}{\partial x_i} - \frac{\partial \Phi}{\partial x_i}\frac{\partial f}{\partial v_i}\right) = 0 \quad (3.11)$$

By integrating equation 3.11 over all possible velocities, we obtain the Jeans equation. For a spherically symmetric system, it is expressed as

$$\frac{1}{\rho}\frac{d(\rho \bar{v}_r^2)}{dr} + 2\frac{\beta \bar{v}_r^2}{r} = -\frac{GM(r)}{r^2} \quad (3.12)$$

where $\rho = \int f d^3\mathbf{v}$ is the spatial density of stars, $\beta = 1 - \bar{v}_\theta^2/\bar{v}_r^2$ is the degree of anisotropy of the velocity distribution and $\bar{v}_i$ is the mean stellar velocity in the given coordinate. For an isotropic velocity distribution, $\bar{v}_r^2 = \sigma^2$ and $\beta = 0$. The mass within the radius r is thus

3.2 Stellar mass estimates

determined by solving the Jeans equation.

Treu & Koopmans ([2004]) have shown that, by combining lensing measurements with spatially resolved kinematic profiles in elliptical galaxies, the stellar and dark matter components could be separated precisely. In particular, this method allows one to break both the degeneracy between mass and velocity anisotropy (for a given velocity dispersion and stellar density, see Eq. 3.12) in stellar dynamics and the so-called "mass-sheet degeneracy" (Falco, Gorenstein & Shapiro [1985], Schneider & Seitz [1995]) which limits the accuracy of lensing methods. The latter is due to the fact that the image of the source is not affected by transformations of the kind $\kappa \rightarrow \kappa' = \lambda\kappa + (1 - \lambda)$ where λ is a constant, i.e. by rescaling the density distribution of the lens and adding a constant density mass sheet. If the velocity dispersion of stars is known only from a single aperture, some information on the stellar mass fraction $f_\star$ inside the Einstein radius R_{Ein} can still be obtained. This particular analysis was performed on the SLACS sample by Koopmans et al. ([2006]) and is summarized here:

- The total mass distribution of each lens galaxy is modeled as a power-law density profile: $\rho(r) = \rho_0 r^{-\gamma}$. The stellar component to the mass is described in terms of a Hernquist ([1990]) profile: $\rho_\star(r) = M_\star r_\star/4\pi r(r+r_\star)^3$, where $r_\star = R_e/1.8153$ and R_e is the effective radius of the galaxy.

- The lensing measurement of the total projected mass enclosed within the Einstein radius R_{Ein} is used to determine ρ_0. Thus, for any given $\{M_\star/L, \gamma\}$, the spherical Jeans equation can be solved to determine the line-of-sight stellar velocity dispersion as a function of radius. This is done assuming different values for the velocity anisotropy β.

- A likelihood function for γ is defined by comparing the predicted and observed velocity dispersions. A prior on the stellar mass-to-light ratio $M_\star/L$, based on the local value of $M_\star/L$ (in the B band) and corrected for the observed evolution of the Fundamental Plane, is adopted before marginalizing on the two free parameters of the model, $M_\star/L$ and γ. If the total density profile was different from a power-law, one would expect the best fit γ inside R_{Ein} to vary depending on where the change in slope occurs with respect to R_e. As Koopmans et al. ([2006]) found no correlation between γ and the ratio R_{Ein}/R_e, they conclude that this assumption is valid.

- The stellar mass fraction $f_\star$ is calculated as the ratio between the maximum likelihood value of $M_\star/L$ and the maximum allowed value of $M_\star/L$. The latter is obtained under the assumption that the stellar mass is equal to the total mass. From the stellar mass fraction $f_\star$, Koopmans et al. derive an estimate for the stellar mass inside the Einstein radius: $M^\star_{len+dyn}(\leq R_{Ein}) = f_\star \times M_{len+dyn}(\leq R_{Ein})$.

3.3 Photometric stellar mass

Unlike strong lensing effects or the central velocity dispersion, the UV to IR luminosity of a passive galaxy is a function of its stellar mass only. In particular, the near-infrared emission is dominated by long-lived stars, which make up the bulk of a galaxy, and insensitive to ongoing star formation. The near-infrared luminosity of a galaxy is therefore a good tracer of its stellar mass (e.g. Gavazzi et al. [1996], Kauffmann & Charlot [1998], Bell & de Jong [2001], Rettura et al. [2006]). Stellar masses can also be estimated from optical colors, as in Bell et al. ([2003]) or van der Wel et al. ([2005]). Finally, as seen in Eq. 2.7, the value of the stellar mass is a straightforward result of fitting the SED of a galaxy with stellar population models (e.g. Rettura et al. [2006], van der Wel et al. [2006]). Here we used this last approach, as it provides further constraints on the mass. We used photometry based on observations of the sample galaxies in the five SDSS bands, u, g, r, i and z at 354, 477, 623, 763 and 913 nm respectively. In order to have unbiased SEDs, we used *modelMag* magnitudes in the publicly available SDSS catalog. These are obtained by fitting a de Vaulcouleurs ([1948]) profile,

$$I(R) = I_0 \, exp\Big(-7.67 \left(\frac{R}{R_e}\right)^{1/4}\Big) \qquad (3.13)$$

where R_e is the standard optical effective radius, to the r-band image of each galaxy and varying the amplitude of the model in the other bands after convolution with the point spread function). The resulting magnitudes correspond to magnitudes measured through equivalent apertures in all bands. The SDSS photometry is corrected for galactic extinction (Adelman-McCarthy et al. [2006]), using the dust maps of Schlegel et al. ([1998]). We derived photometric stellar masses $M^\star_{phot}$ for each galaxy in the sample by fitting its observed SED with τ-models computed from single stellar population templates at solar metallicity. We considered T values of 200 Myr to the age of the Universe at the galaxy's redshift with increments of $\sim$250 Myr, τ values of 0 to 1 Gyr with increments of 50 Myr and values of $M^\star_{phot}$ between 10^9 and 10^{13} $M_\odot$ with increments of $0.1 \lfloor \log(M) \rfloor$ dex. We used Bruzual & Charlot ([2003]) templates with either a Salpeter ([1955]) or Chabrier ([2003]) initial mass function as well as Maraston ([2005]) templates with a Salpeter or a Kroupa ([2001]) IMF.

The photometric stellar mass inside the Einstein radius $M^\star_{phot}(\leq R_{Ein})$ was then inferred by multiplying the mass $M^\star_{phot}$ derived from the SED fit by an aperture factor

$$f_{ap} = \frac{\int_0^{R_{Ein}} I(R) R dR}{\int_0^\infty I(R) R dR} \qquad (3.14)$$

which represents the fraction of light enclosed inside the Einstein radius with respect to the total light of the galaxy parametrized by the de Vaucouleurs profile, with the implicit assumption that the stellar mass is traced by the light distribution. Fig. 3.2 shows the SED and best fit model of six galaxies in the sample. From the description of both stellar mass estimates, $M^\star_{phot}(\leq R_{Ein})$ and $M^\star_{len+dyn}(\leq R_{Ein})$, it is clear that these are two independent measurements of the stellar mass. It is therefore interesting to compare them.

3.4 Comparison between the different stellar mass estimates

In Fig. 3.3, we compared the photometric stellar masses obtained from fitting the SEDs of the galaxies in the SLACS sample with the masses of Koopmans et al. ([2006]) obtained from lensing and stellar dynamics. The plots show that the masses derived from models with a Salpeter IMF are consistent, within the error bars, with the Koopmans et al. estimates. The best fit correlation line yields

$$M^\star_{len+dyn}(\leq R_{Ein}) = 10^{2.3\pm1.68} \times M^\star_{phot}(\leq R_{Ein})^{0.8\pm0.15} \quad (3.15)$$

with a correlation coefficient of $\rho = 0.94$. The median value of the ratio $q = M^\star_{len+dyn}/M^\star_{phot}$ between the lensing+dynamical and photometric mass estimates is consistent with unity (1.1±0.1) but does not show any correlation with galaxy colors, thus excluding a possible source of systematic errors in the photometric mass estimates. No significant difference was found when using the Maraston ([2005]) templates over the Bruzual & Charlot ([2003]) ones with the same IMF. This is not surprising as the two population synthesis models differ remarkably only for ages lower than 2 Gyr, i.e. for younger populations than the ones examined in this Chapter. This result is in agreement with that of Rettura et al. ([2006]). Values of q larger than one may be explained by possible underestimates of $M^\star_{phot}(\leq R_{Ein})$. These can be ascribed to two different phenomena: dust extinction and metallicity values lower than solar. Both effects would tend to produce lower fluxes with respect to a solar metallicity or dust-free model. This would naturally result in lower mass estimates. By choosing a more top-heavy IMF like Kroupa ([2001]) or Chabrier ([2003]), the photometric mass estimates were lowered in such a way that the slope of the best fit was unchanged but the q value was considerably larger than one.

As an additional check, we investigated possible biases due to the source galaxy. Since the image of the source is very close to the center of the lens and much bluer (see Fig. 3.1), a combined SED would appear bluer and be best fitted with younger models, resulting in an underestimate of $M^\star_{phot}(\leq R_{Ein})$. To quantify this, we compared the photometric stellar masses of the SLACS and SLACS II lens galaxies estimated from the fit on their SEDs, alternatively without the u or z band. As shown in Fig. 3.4, we found no significant difference between the mass estimates from the full 5-band SEDs, the masses derived from the g, r, i and z bands and those from the u, g, r and i bands. This shows that the SED fit is robust and that contamination of the $modelMag$ magnitudes due to the source image is negligible.

Finally, we measured the stellar mass-to-light ratio $M_\star/L_B$ of the galaxies in the SLACS and SLACS II samples using the photometric stellar mass and the rest-frame B-band flux

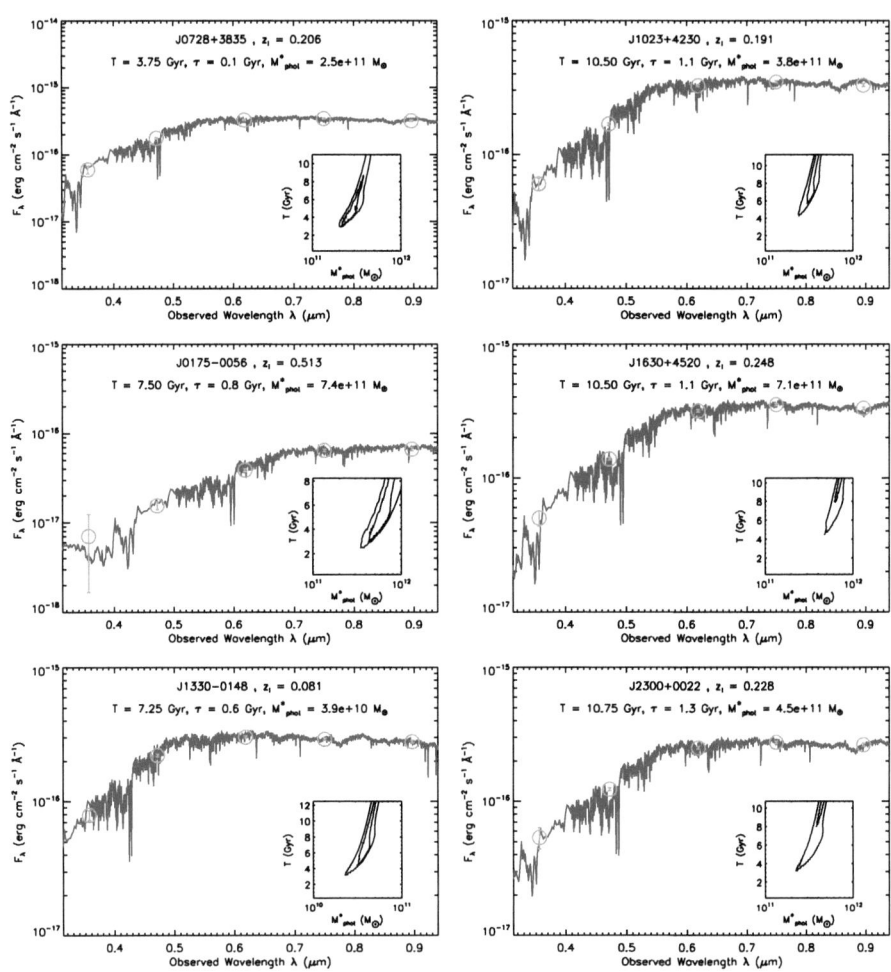

Figure 3.2: SEDs and best fit models of the lens galaxies SDSS J0728+3835, SDSS J1023+4230, SDSS J0175-0056, SDSS J1630+4520, SDSS J1330-0148 and SDSS J2300+0022, at $z = 0.206, 0.191, 0.513, 0.248, 0.081$ and 0.229 respectively. The red circles with error bars show the observed total flux densities measured in the u, g, r, i, z SDSS passbands. The best fit model is shown in blue. On the bottom right of each plot, the insets show the 1σ and 3σ confidence regions for T and $M^\star_{phot}$.

3.4 Comparison between the different stellar mass estimates

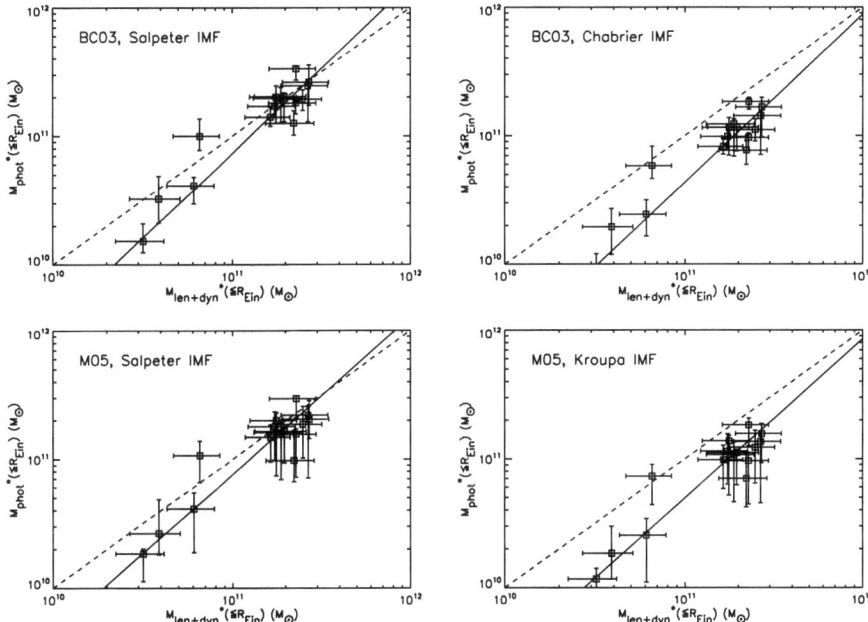

Figure 3.3: Comparison of lensing+dynamical and photometric stellar masses measured inside the Einstein radii of the SLACS sample of gravitational lens early-type galaxies, for photometric masses estimated using BC03 (*top*) and M05 (*bottom*) models and assuming a Salpeter (*left*) or more top-heavy (*right*) IMF. The solid line shows the best fit correlation while the dotted line corresponds to $M^\star_{len+dyn}(\leq R_{Ein}) = M^\star_{phot}(\leq R_{Ein})$. The photometric stellar masses obtained with a Salpeter IMF are consistent with the lensing+dynamical ones, while the photometric stellar mass estimates obtained using a top-heavy IMF are significantly lower.

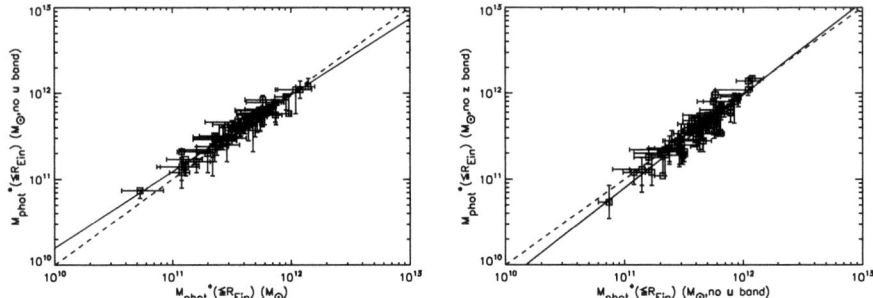

Figure 3.4: Left : comparison between the photometric stellar masses obtained from fitting the 5-band SEDs of galaxies in the SLACS and SLACS II samples and those derived from fitting the 4-band SEDs without the u band. Right : comparison between the photometric stellar masses of galaxies in the SLACS and SLACS II samples derived from fitting their 4-band SEDs, without the u or the z bands. The three photometric mass estimates are consistent with one another, showing the robustness of the SED fit.

predicted by the best fit models. As the rest-frame B band lies between the SDSS filters at the redshift of the galaxies in the sample, this latter value is only weakly model-dependent. The photometric stellar masses (and related uncertainties) of the SLACS II galaxies were computed as described above. The stellar mass-to-light ratios were then compared to the values expected from the evolution of the Fundamental Plane. As shown in Fig. 3.5, the mass-to-light ratios derived from models with a top-heavy IMF (Kroupa or Chabrier) are systematically smaller than those predicted by the Fundamental Plane, while the Fundamental Plane values are consistent with those derived from models with a Salpeter IMF, within errors.

This analysis thus showed that photometric mass measurements obtained by choosing a solar metallicity model with a Salpeter IMF are reliable. Furthermore, this makes the presence of strong biases in one of the two methods very unlikely, allowing them to be used independently to reliably measure stellar masses. The underestimate of the photometric stellar mass with a Chabrier or Kroupa IMFs, with respect to a Salpeter IMF, is due to their higher proportion of solar and super-solar mass stars, which dominate the total light of the galaxy at these ages. This leads to model SEDs with higher fluxes, which therefore require less stellar mass to fit the observed SED. Therefore, the relation between the different mass estimates depends on the mass cut-offs and a good agreement between the lensing+dynamical and photometric stellar masses could also be reached by adopting a lower mass cut-off at the low end of the IMF. We note however that, for the top-heavy IMFs to have roughly the same amount of solar and super-solar mass stars as the Salpeter IMF, the low mass cut-off would have to be lower than the hydrogen-burning mass limit of

3.4 Comparison between the different stellar mass estimates

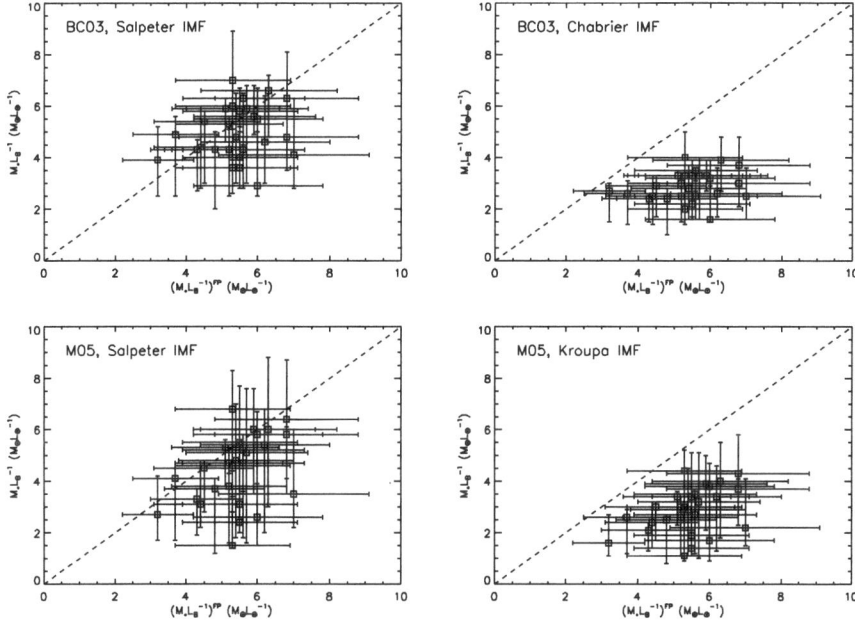

Figure 3.5: Comparison of the mass-to-light ratios predicted by the evolution of the fundamental plane and those derived from photometric stellar mass estimates, using BC03 (*top*) and M05 (*bottom*) models and assuming a (*left*) or more top-heavy (*right*) IMF. The dashed line shows the 1:1 correlation. As in Fig. 3.3, the estimates derived with a top-heavy IMF are systematically lower than the predictions.

$0.08 M_\odot$, leading to an unphysical scenario. Although this study was based on a relatively low redshift sample, the photometric mass estimates can be expected to be also accurate to higher redshifts as long as the same optical/near-IR rest frame bands are covered, which is easily achievable these days with JHK photometry, e.g. from the ISAAC instrument at the VLT observatory or SofI at the New Technology Telescope (NTT), or with IRAC (3.5 to 8 μm) on the Spitzer space observatory.

3.5 Visible and dark matter

In a second approach, we compared the total (lensing) masses and luminous (photometric stellar) masses, both within the Einstein radius, of 27 galaxies in the SLACS and SLACS II samples (see Tables 3.1 and 3.2). The total (luminous and dark matter) masses within the Einstein radius and their uncertainties were obtained using Eq. 3.6, assuming a 5% error on the Einstein angle. This allowed us to study the relationship between luminous and dark matter as a function of the Einstein radius and mass. Fig. 3.6 shows the total mass estimates from strong lensing versus the photometric mass estimates from SED fitting for different models and initial mass functions. We note that the best fit correlation line of all the models is roughly parallel to the one-to-one correlation line, with the photometric masses derived from templates with a top-heavy IMF being significantly lower than the photometric masses obtained with a Salpeter IMF. Specifically, we found that the mean value of the ratios between the total and luminous masses was 1.5±0.1 and 1.7±0.1 in the case of a Salpeter IMF (for BC03 and M05 templates respectively), 2.5±0.2 and 2.6±0.2 for top-heavy IMFs (for BC03 and M05 templates respectively). This suggests that the total mass is proportional to the luminous mass of the lenses in our sample and that, on average, the amount dark matter that must be added to the luminous matter to give the total mass of the lenses is higher for the models computed with top-heavy IMFs with respect to those with a Salpeter IMF.

We then evaluated the fraction of mass in the form of stars within the Einstein radius, $f_\star(\leq R_{Ein})$, as the ratio between the photometric stellar mass estimates and the total mass obtained from lensing. Fig. 3.7 shows the stellar mass fraction as a function of the adimensional ratio of the Einstein radius to the effective radius of each lens, which is roughly independent of the redshift of the lens and thus describes better than the sole Einstein radius how close the investigated region is to the center of the lens the (Koopmans et al. [2006]). Fig. 3.8 shows the same points binned in three intervals $R_{Ein}/R_e \leq 0.5$, $0.5 < R_{Ein}/R_e \leq 0.65$ and $R_{Ein}/R_e \geq 0.65$. Fig. 3.9 shows the stellar mass fraction as a function of the redshift of the lens galaxy, for the three redshift bins $z_l \leq 0.17$, $0.1 < z_l \leq 0.25$ and $z_l > 0.25$. We found that $f_\star(\leq R_{Ein})$ stays roughly constant from 0.2 to 1 R_e. A similar trend was also found in the analysis of the SLACS sample performed by Koopmans et al. ([2006]). The stellar mass fraction seems to increase at first and then decrease around 0.8 R_e, but we do not consider it significant, as the signal is low and it could be an artifact of the total mass model (see Section 3.2). Fig. 3.9 suggests

3.5 Visible and dark matter

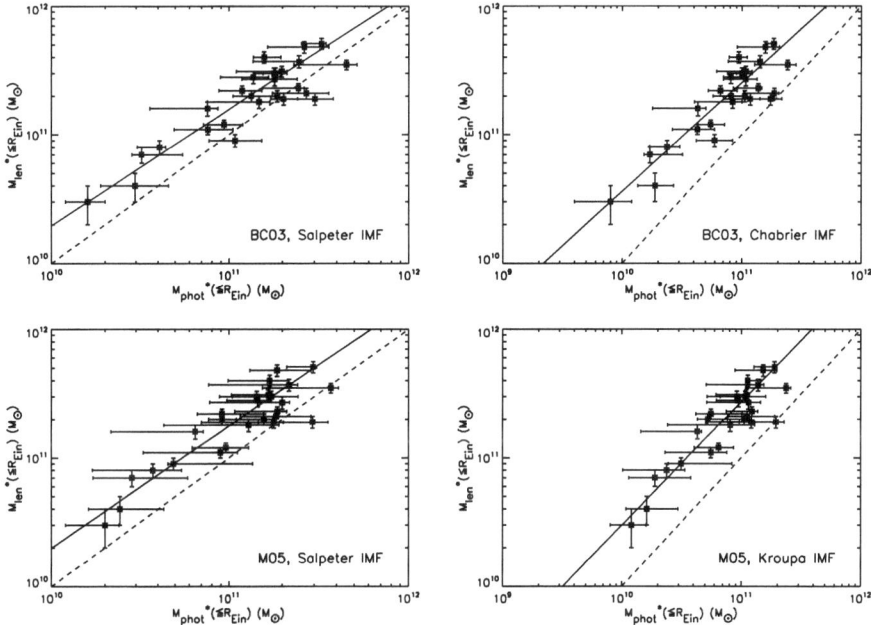

Figure 3.6: Total mass derived from strong gravitational lensing as a function of the photometric stellar mass within the Einstein radius, estimated by SED fitting using BC03 (*top*) and M05 (*bottom*) models and assuming a Salpeter (*left*) or more top-heavy (*right*) IMF. The best fit correlation and the one-to-one relation are shown by solid and dotted lines respectively.

that this is not due to possible evolutionary effects, that would increase the dark matter concentration in the central regions of the lens galaxies over time. For elliptical galaxies, whose stellar content varies little, this would result in a decrease of $f_\star(\leq R_{Ein})$ with age, i.e. with increasing redshift.

Fig. 3.7 and Fig. 3.8 show that a significant (more than 30%) amount of dark matter is concentrated in the inner part ($\lesssim 0.2 R_e$) of the lens elliptical galaxies. Furthermore, the slope of the cumulative mass profile of the dark matter is equal to or shallower than that of the luminous matter up to at least $\sim 0.8 R_e$, resulting in a constant (or perhaps even increasing) stellar mass fraction. Beyond this value, the distribution of the dark matter density may overcome that of the luminous one, but additional measurements of lens galaxies with $R_{Ein} \geq 0.8 R_e$ would be needed to verify this.

3.6 Summary

In this Chapter, we have verified that the stellar masses estimates of elliptical galaxies obtained by comparing their spectral energy distributions to stellar population synthesis models were well consistent, within uncertainties, with estimates based on a combination of strong gravitational lensing and stellar dynamics. As these methods are completely independent, this makes the presence of strong biases in one of the two very unlikely. This means that they can be used independently to reliably estimate stellar masses. The sample used in this Chapter is a relatively low-redshift one, only extending to $z \sim 0.5$, but we can expect photometric mass estimates to be also accurate at high-redshift, at least for early-type galaxies, as long as the observations cover the same optical and near-IR rest-frame bands. We found that photometric stellar mass estimates were not model-dependent, as the ages of the elliptical galaxies in the SLACS and SLACS II samples were not in the range where the two sets of stellar population models used differ. We also found that the photometric stellar mass estimates vary strongly with the adopted initial mass function. For the same mass cut-offs of $0.1 M_\odot$ and 100 $M_\odot$ and assuming solar metallicity, we found that best fit models using a Salpeter IMF were more consistent with the combined estimates from lensing and stellar dynamics, and reproduced the expected mass-to-light ratios better, than best fit models using a more top-heavy IMF such as Chabrier ([2003]) or Kroupa ([2001]). For this reason, templates with a Salpeter IMF with mass cut-offs at 0.1 $M_\odot$ and 100 $M_\odot$ are used in the following Chapters. This result allowed us to compare the luminous and total masses of a broader sample of lens elliptical galaxies. From this test, we found that the total mass is proportional to the luminous (photometric) mass and that a large ($\gtrsim$30%) amount of dark matter appears to be present in the central regions of the lens galaxies, in accord with previous studies of dark matter in elliptical galaxies using stellar dynamics (Saglia, Bertin & Stiavelli [1992], Ortwin et al. [2001]) and X-ray emission (Mushotzky et al. [1994], Loewenstein & White [1999]). Furthermore, the stellar mass fraction appears nearly constant up to $\sim 1 R_e$, suggesting that the profile of the dark matter component is similar to that of the stellar one within this region.

3.6 Summary

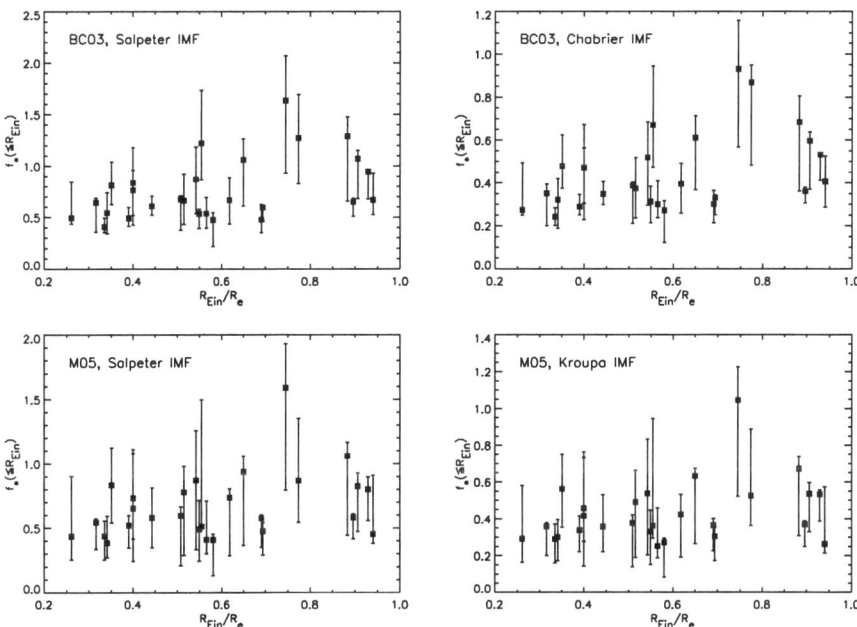

Figure 3.7: Fraction of mass in the form of stars $f_\star$ enclosed within the disk defined by the Einstein radius as a function of the Einstein radius (in units of the effective radius) of the 27 galaxies in the SLACS and SLACS II samples. The photometric mass estimates were obtained using Bruzual & Charlot (BC03, top) and Maraston (M05, bottom) models with Salpeter (left) and Chabrier/Kroupa (right) initial mass functions. The constant $f_\star$ up to at least $0.8 R_e$ implies that the dark matter profile is equal or shallower to that of the luminous matter.

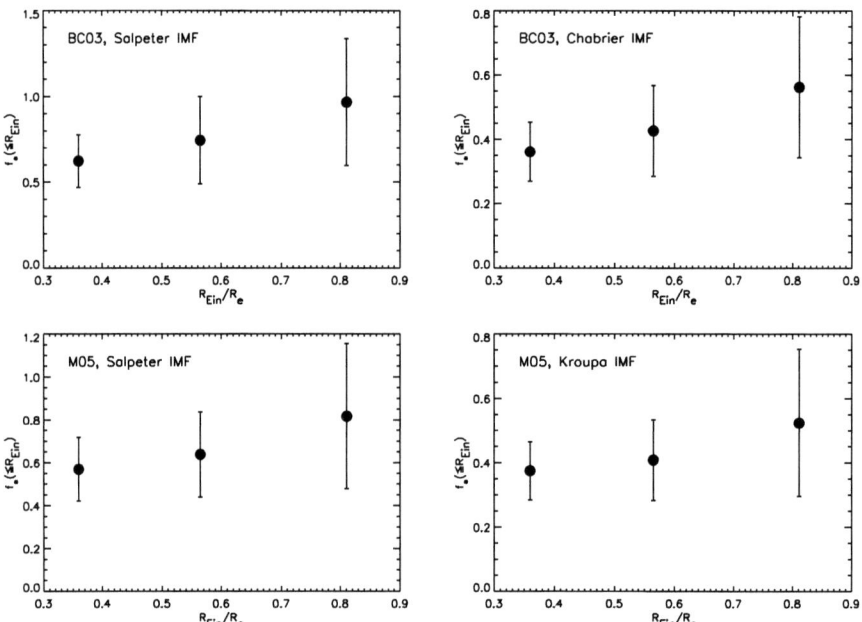

Figure 3.8: Fraction of mass in the form of stars enclosed within the Einstein radius as a function of the Einstein radius (in units of the effective radius) of the 27 galaxies in the SLACS and SLACS II samples, for BC03 (top) or M05 (bottom) models with Salpeter (left) or Chabrier/Kroupa (right) IMFs. The same points of Fig. 3.7 were binned in three intervals of Einstein radius: $R_{Ein} \leq 0.5R_e$, $0.5R_e < R_{Ein} \leq 0.65R_e$ and $R_{ein} > 0.65R_e$.

3.6 Summary

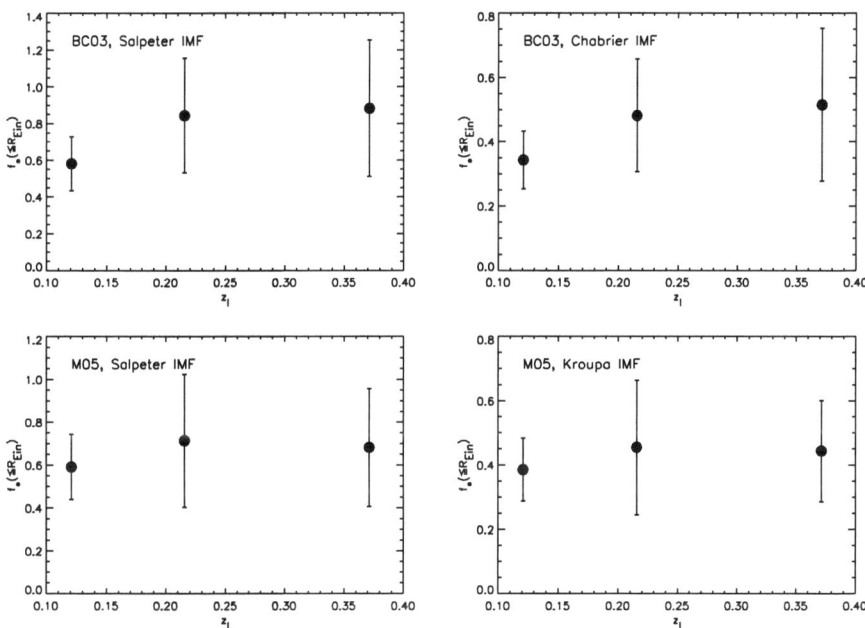

Figure 3.9: Fraction of mass in the form of stars enclosed within the Einstein radius as a function of the redshift of the lenses, binned in the three intervals $z_l \leq 0.17$, $0.1 < z_l \leq 0.25$ and $z_l > 0.25$. The templates used for the models were BC03 (top) and M05 (bottom) and the initial mass functions Salpeter (left) and Chabrier or Kroupa (right).

Chapter 4

Star formation histories in cluster and field at $z \sim 1.2$

As stated in Chapter 1, massive early-type galaxies are a good tracer of the early mass assembly in the Universe. The study of their spectroscopic, photometric and morphological properties, especially at high redshift, over a range of environmental densities can significantly constrain the different models of structure formation and bring insight to the physical mechanisms of galaxy formation and evolution. Hierarchical galaxy formation (e.g. Toomre [1977]), in particular, is naturally expected in a ΛCDM cosmology and predicts different formation histories whether a galaxy is in a low-density environment or member of a cluster (e.g. De Lucia et al. [2006]). Indeed, the analysis of the fossil record via line-strength indices shows that massive early-type galaxies in local high-density environments are at least 1.5 Gyr older than their counterparts in low-density regions (Thomas et al. [2005], Sánchez- Blázquez et al. [2006], Clemens et al. [2006]), while from the mass-to-light ratio of cluster and field galaxies up to $z \sim 1$, van Dokkum & van der Marel ([2007]) found a lower value of ~ 0.4 Gyr. On the other hand, early-type galaxies appear to have formed at an early time and been in place at $z \simeq 2$ (e.g. Bernardi et al. [1998], van Dokkum et al. [2001a]), with little star formation happening ever since. This suggests that studying the star formation history of early-type galaxies at $z > 1$ allows one to place stronger constraints on formation and evolution models than at low redshift, where any difference has been smoothed out by billions of years of mostly passive evolution. In such a case, much sparser data are available, as few massive galaxy clusters have been confirmed so far at redshifts greater than unity. One of these clusters, RDCS J1252.9-2927 at $z = 1.237$, was discovered as part of the ROSAT Deep Cluster Survey (Rosati et al. [1998], [2004]). Near-infrared follow-up observations showed an overdensity of red ($J - K_s \sim 1.85$) galaxies consistent with a population of early-type galaxies at $z > 1$ (Lidman et al. [2004]). Subsequently, RDCS J1252.9-2927 was the target of an extensive multi-wavelength observation campaign (Blakeslee et al. [2003]), Lidman et al. [2004], Demarco et al. [2007]). In this chapter, we use the spectroscopic and photometric data of early-type galaxies in RDCS J1252.9-2927 to reconstruct their general star formation history and compare it with early-type galaxies selected at similar redshift from the Great Observatories Origin Deep

Survey/Chandra Deep Field-South (GOODS/CDF-S for short, Dickinson et al. [2003]). This Chapter is organized as follows. In Section 4.1, we describe our cluster and field samples as well as our selection criteria. In Section 4.2, we describe the grid of composite stellar population models used to fit the spectrophotometric data of our samples. In Section 4.3, we present the results of the spectrophotometric analysis and discuss its possible biases. In Section 4.4, we compare the results of our spectrophotometric analysis to the prediction of a semi analytic model of galaxy formation and evolution. Finally, in Section 4.5, we use our results to estimate the scatter of the red sequence up to $z \sim 2$.

4.1 Data and sample selection

We used RDCS J1252.9-2927 observations in 9 bands in the 0.4-5 μm range, namely the B, V and R bands (Demarco et al. [2007]) with FORS2 on the VLT, F775W and F850LP bands (hereafter i_{775} and z_{850}, Blakeslee et al. [2003]) with the ACS Wide Field Camera on the HST, the J_s and K_s (Lidman et al. [2004]) with ISAAC (Moorwood [1992]) on the VLT and at 3.6 and 4.5 μm with IRAC (Fazio et al. [1998]) on the Spitzer space telescope. A color composite image of the 2'×2' central region of RDCS J1252.9-2927 is shown in Fig. 4.1. The magnitudes were measured using SExtractor (Bertin & Arnouts [1996]) in 1.5" diameter apertures and then corrected to 2". The photometry was corrected for galactic extinction using the NASA/IPAC Extragalactic Database (NED[1]) tool, which uses the dust maps of Schlegel et al. ([1998]). As Schlegel et al. used the SED of an elliptical galaxy to calculate the extinction corrections, these are appropriate for a sample of early-type galaxies. The extinction corrections are 0.327 mag in B, 0.242 in V, 0.198 in R, 0.157 in i_{775}, 0.121 in z_{850}, 0.071 in J_s, 0.029 in K_s and 0.015 and 0.013 in the two IRAC bands respectively. The AB corrections to the FORS2, ACS and ISAAC magnitudes were -0.088, 0.052, 0.244, 0.401, 0.569, 0.964 and 1.899 respectively. In addition, an extensive spectroscopic campaign of this cluster was carried out with VLT/FORS2 using the 300I grism (Demarco et al. [2007]), which provides a resolution of about 13$\mathring{A}$ at 8600 $\mathring{A}$.

As a low-density region, we chose the GOODS/CDF-S field for its deep and extensive photometric and spectroscopic coverage. The photometric data for the GOODS/CDF-S field comprises observations in the F435W, F606W (hereafter b_{435} and v_{606}), i_{775} and z_{850} bands with HST/ACS (Giavalisco et al. [2004]), the J and K_s bands with VLT/ISAAC and at 3.6 and 4.5 μm with $Spitzer$/IRAC. As for RDCS J1252.9-2927, we used magnitudes measured in 1.5" diameter corrected to 2" diameter. In addition, we used spectra taken using VLT/FORS2 with the 300I grism (Vanzella et al. [2005], [2006], [2008]).

As a result, the two datasets used in this analysis have similar quality and wavelength coverage, thus yielding homogeneous spectrophotometric properties of galaxies and allowing a single selection criterion for both samples. Specifically, the availability of 8 to 9 passbands allowed the estimate of reliable photometric stellar masses (Rettura et al. [2006]).

---
[1] http://nedwww.ipac.caltech.edu/

4.1 Data and sample selection

Figure 4.1: 2'×2' ($\sim$ 1 Mpc at $z = 1.24$) region centered on RDCS J1252.9-2927, from a color composite of VLT/FORS2 (B), HST/ACS (z_{850}) and VLT/ISAAC (K_s) images (Rosati et al. [2004]).

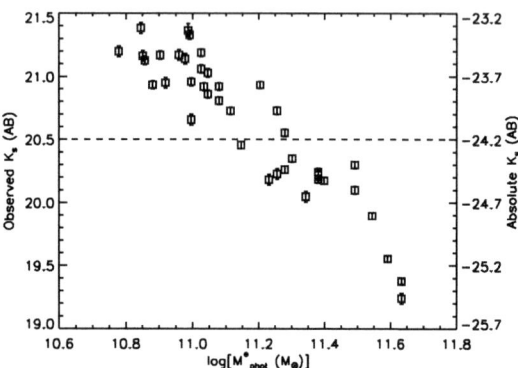

Figure 4.2: Stellar mass as a function of K-band magnitude: photometric stellar masses of red ($i_{775} - z_{850} \geq 0.8$) galaxies at $z \sim 1.24$ in RDCS 1252 and GOODS as a function of their K_s-band magnitude. The dashed line shows the $K_s^\star$ value of the best Schechter ([1976]) fit to the luminosity function of cluster galaxies at $z \sim 1.2$, obtained by Strazzullo et al. ([2006]).

We derived photometric stellar masses by SED fitting with composite stellar population models, as described in Chapter 3. These were computed assuming solar metallicity and a Salpeter ([1955]) initial mass function, using the Bruzual & Charlot ([2003]) templates. The near-IR depth of both datasets allowed us to define complete mass-selected samples. The GOODS/CDF-S and RDCS J1252.9-2927 (hereafter GOODS and RDCS 1252 for brevity) K_s-band images are photometrically complete down to at least $K_s = 24$. At $z \simeq 1.2$, the K_s-band photometry is a very good proxy of the stellar mass (e.g. Kauffmann & Charlot [1998], Rettura et al. [2006]), as shown in Fig. 4.2, with $10^{10} M_\odot$ corresponding to $K_s \simeq 23$ (Strazzullo et al. [2006]), which we took as a reliable mass completeness limit. On the other hand, the spectroscopic follow-up work is limited to $K_s \simeq 22$ for early-type galaxies in both samples, corresponding to $R_{Vega} \simeq 25$ and $M_\star \simeq 3 \times 10^{10} M_\odot$. Therefore, we limited our analysis to photometric stellar masses greater than $M_{lim} = 5 \times 10^{10} M_\odot$.

The selection of early-type galaxies in the cluster is apparent from the well-defined red sequence (Blakeslee et al. [2003]) which clearly lies above $i_{775} - z_{850} = 0.8$ (see Fig. 4.3) and thus separates the population of blue star-forming galaxies at the cluster redshift. Spectra were available for 22 of the 38 galaxies on the red sequence with $M_\star \geq M_{lim}$. For the corresponding GOODS field sample, we selected galaxies with $M_\star \geq M_{lim}$, in the redshift range $z = 1.237 \pm 0.15$ and adopted the same $i_{775} - z_{850} \geq 0.8$ cut used in RDCS 1252. This criterion yielded 21 early-type galaxies with FORS2 spectra. We verified that the spectra did not show signs of ongoing star formation, i.e. that no [OII]$\lambda 3727$ emission line was detectable. We assumed that the galaxies in our samples are dust-free (see Section

4.1 Data and sample selection

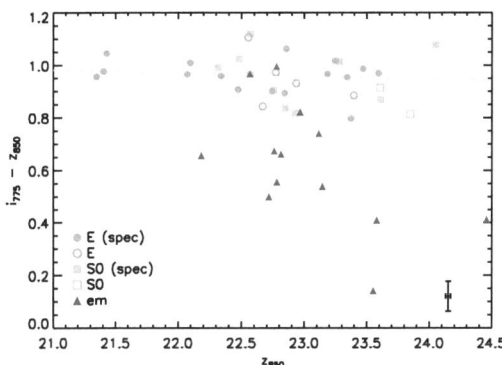

Figure 4.3: $i_{775} - z_{850}$ color-magnitude relation of early-type galaxies in RDCS J1252.9-2927 with $i_{775} - z_{850} \geq 0.8$ and $z_{850} \leq 24$. The circles and squares represent elliptical and S0 galaxies, respectively, while the triangles show emission line galaxies. The filled symbols correspond to spectroscopically confirmed members. The dashed line show the best fit linear correlation to the E and S0 galaxies.

4.3, also Rettura et al. [2006]). The vast majority of the galaxies in both samples have E/S0 morphologies following the classification scheme used by Blakeslee et al. ([2003]). Of the 18 galaxies in the GOODS sample that are found in the GEMS (Häussler et al. [2007]) catalog, all have a Sérsic index greater than 3. Fig. 4.4 shows the spatial distribution of the early-type galaxies in the RDCS 1252 sample in the cluster's field. Spectroscopic errors were based on the evaluation of the signal-to-noise ratio at 4100 Å rest-frame from the residuals after fitting the H_δ absorption line, as described in Chapter 2. The total response of the detector was then normalized to this value to obtain a noise estimate as a function of wavelength.

An important aspect to be addressed is the relative spectroscopic completeness as a function of mass of the GOODS and RDCS 1252 samples, which should be considered when interpreting the results from our comparative spectrophotometric analysis. This can be quantified with accurate photometric redshifts available for both samples, which were derived using the Coleman et al. ([1980]) templates and the BPZ code (Benítez [2000]), as described in Toft et al. ([2004]). Using photometric redshifts, we found 117 GOODS galaxies in the given redshift bin, with $i_{775} - z_{850} \geq 0.8$ and $M_\star \geq M_{lim}$. In Fig. 4.5, we compared the photometric mass distributions of the color- and mass-selected galaxies in GOODS and RDCS 1252 to the mass distribution of the spectroscopically observed subsamples. It is immediately apparent that the spectroscopic follow-up work was more extensive in RDCS 1252 than in GOODS; as a result the GOODS spectroscopic sample is more incomplete at the low mass end than the RDCS 1252 sample (table 4.1). This point

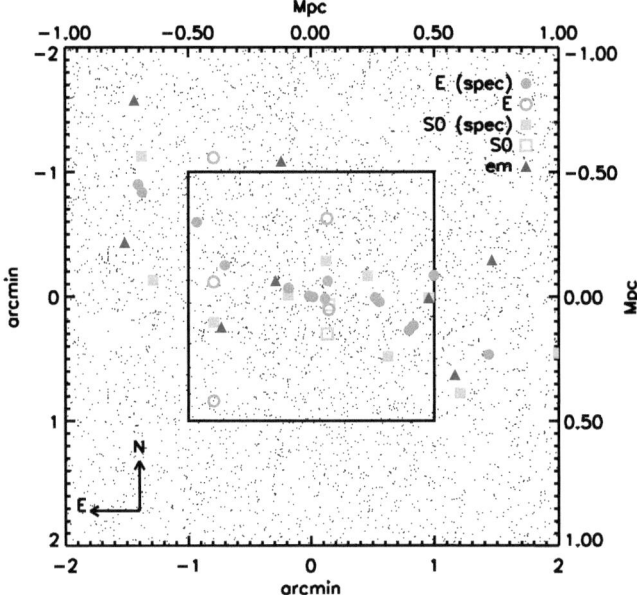

Figure 4.4: 4'×4' map centered on RDCS J1252.9-2927 and spatial distribution of the red sequence galaxies. The circles and squares represent elliptical and S0 galaxies, respectively, while the triangles show emission line galaxies. The filled symbols correspond to spectroscopically confirmed members. The square box corresponds to 2'×2', or $\sim 1 \times 1$ Mpc. The field is centered at 12h 52m 54.4s and -29° 27' 17".

4.2 Spectrophotometric modeling

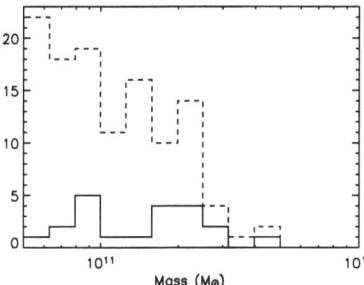
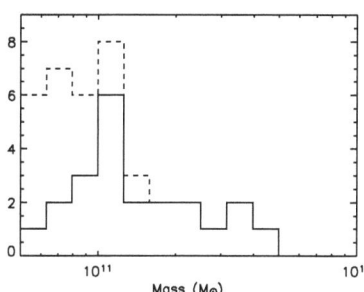

Figure 4.5: Mass completeness of the GOODS and RDCS 1252 samples: distribution of galaxy masses in the GOODS (left) and RDCS 1252 (right) samples. The dashed line shows the photometric sample and the solid line the spectroscopic sample.

sample	Lower mass cutoff ($\times 10^{10} M_\odot$)			
	5	9	15	23
GOODS	18%	25%	35%	67%
RDCS 1252	59%	83%	100%	100%

Table 4.1: Cumulative spectroscopic completeness of the GOODS and RDCS 1252 samples.

is discussed further in Section 4.3.

4.2 Spectrophotometric modeling

We compared the spectrophotometric data of the cluster and field samples to CSP models generated using the Bruzual & Charlot ([2003]) templates, using the method described in Section 2. For this analysis, we assumed a delayed, exponentially declining star formation history parametrized by the time-scale τ, as described in Chapter 2. Since we had 8-9 photometric bands as well as spectroscopic features, we allowed for a more complex star formation history and expanded the grid of models by adding a secondary episode of star formation after the main event, parametrized by an instantaneous burst at time $t_{burst} > \tau$ of amplitude A, so that the star formation rate is expressed as :

$$\Psi(t) = \tau^{-2} t e^{\frac{-t}{\tau}} + A\delta(t - t_{burst}) \quad (4.1)$$

We considered age values, i.e. the time T after the onset of star formation, ranging from 200 Myr to 5 Gyr (about the age of the Universe at $z \sim 1.2$) in increments of ~ 250 Myr and τ values from 0 (corresponding to a simple, instantaneous burst) to 1 Gyr, in increments of 0.1 Gyr. In the case of a secondary burst, t_{burst} ranged from 1 to 4 Gyr in increments of 1 Gyr with amplitudes A of 0.1, 0.2 and 0.5, which correspond to 1/11, 1/6 and 1/3 of the final stellar mass respectively. Thus, the grid $\{T, \tau, t_{burst}, A\}$ used in this analysis contains $20 \times 13 \times 4 \times 3 + 13 \times 20 = 3380$ models. For the main analysis, all models were computed at solar metallicity and without dust. This assumption is discussed in Section 4.3. The choice of the initial mass function (in the case of the Bruzual & Charlot templates, Salpeter or Chabrier) has little effect on the shape of the composite spectrum for the considered age range ($T \leq 5$ Gyr), as the Salpeter and Chabrier IMFs are nearly identical above 1 $M_\odot$. Nonetheless, we assumed a Salpeter IMF for consistency with the photometric masses described above.

The model spectra were smoothed to match the resolution of the GOODS and RDCS 1252 samples and cropped to a 500 Å interval (rest-frame) centered on the 4000 Å break, which roughly corresponds to the high S/N, well flux-calibrated part of the spectra in our samples. By using this relatively narrow wavelength interval, possible distortions due to uncertain flux calibration on the red end of the spectra (Demarco et al. [2007]) are also minimized. Additionally, the spectra in the RDCS 1252 sample were corrected using very high S/N spectra of four bright cluster galaxies of RDCS 1252, taken by Holden et al. ([2006]).

Because the signal-to-noise ratio of the available FORS2 spectra (ranging from 2 to 6 at 8000 Å for RDCS 1252) does not add more constraints on the star formation histories than those given by the broad-band photometry (see Chapter 2), we stacked the the spectra of early-type galaxies in each sample, in a pixel-by-pixel basis with a 1-σ rejection criterion and without iteration. Fig. 4.6 shows the composite (stacked) spectrum of the 10 and 20 K_s-brightest galaxies in RDCS 1252. The stacked spectrum of the GOODS sample has a S/N of approximately 44 and an equivalent exposure time of $t_{exp} = 93h$. For the RDCS 1252 sample, the corresponding values are S/N~20 and $t_{exp} = 74h$. Likewise, we constructed for each sample an average SED of the early-type galaxies by co-adding their individual SEDs.

4.3 Results

In Fig. 4.7, we show the confidence regions of the fit to the stacked photometric and spectroscopic data of the GOODS and RDCS 1252 samples. In Fig. 4.8, we show the averaged SED and spectrum of galaxies in the GOODS and RDCS 1252 samples compared with the models within the 3σ confidence regions of the fit to the data. As described in Chapter 2, we used the star formation weighted age t_{SFR} and the final formation time t_{fin} to characterize the star formation histories of the best fitting models. Fig. 4.9, left, shows the distribution, weighted by the χ^2, of the look-back time $T - t_{fin}$ from $z = 1.24$ to the final formation time of the best fitting models which lie within the intersection of the 3σ confidence regions of the fit to the photometric and spectral data of the GOODS field and RDCS 1252 cluster samples. The top axis gives the corresponding final formation redshift z_{fin}. Fig. 4.9, right, shows the χ^2-weighted distribution of $T - t_{fin}$ of models within the 3σ confidence region of the fit on the photometric data only. Likewise, Fig. 4.10 shows the

4.3 Results

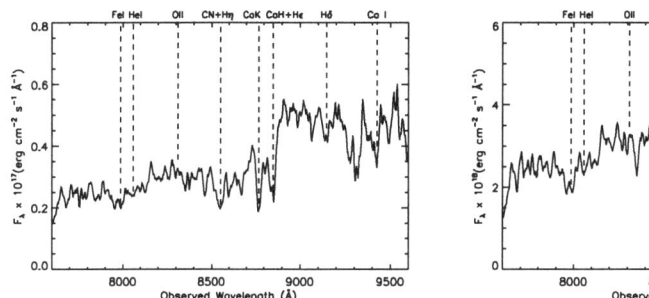

Figure 4.6: Composite spectrum of the 10 (left) and 20 (right) K_s-brightest passive members of RDCS J1252.9-2927. The spectra have been smoothed by 5 pixels ($\sim 12.5\,\text{Å}$). The important spectral features are indicated by dashed lines.

χ^2-weighted distribution of the star formation weighted ages of best fitting models within the 3σ confidence regions of the fit to the spectrophotometric (left) and photometric (right) data of the field and cluster samples. Fig. 4.11 shows the distribution of the ages T of the best fitting models.
Fig. 4.12 shows the median stellar mass fraction $m_\star(t)$, formed at time t after the onset of star formation, in cluster and field galaxies. This is computed as the median of the integrated star formation rate (Eq. 4.1) of the best fitting models. The error bars represent the standard deviation of the distribution of $m_\star(t)$.

From Fig. 4.11 we see that there is no significant delay between environments in the start of the star formation, as the distributions of T are very similar. However, we found that the mean final formation time of the cluster early-type galaxy population is $\sim$1 Gyr greater than the final formation time of the corresponding population in the field, i.e. that field early-type galaxies at $z \sim 1.2$ have longer star formation time scales. The mean residual star formation at $z \sim 1.2$ of the best fitting models is 0.4 $M_\odot yr^{-1}$ for the GOODS sample and 0.06 $M_\odot yr^{-1}$ for RDCS 1252. Likewise, the star formation weighted ages of the two populations differ by $\sim$0.5 Gyr on average with cluster early-type galaxies forming at $z \sim 4$ and field early-types at $z \sim 3.2$. This age difference is in very good agreement with that derived by van Dokkum & van der Marel ([2007]) from the evolution of the mass-to-light ratio, based on a completely independent method and data set. Interestingly, the (χ^2-weighted) distributions of models fitting only the averaged SEDs of the cluster and field samples are very similar, with a mean age difference of less than 0.1 Gyr, meaning that the spectroscopic data carries most of the weight for the difference in time scales. Table 4.2 summarizes the results.

This greater sensitivity of the spectra with respect to the photometric data can be appreci-

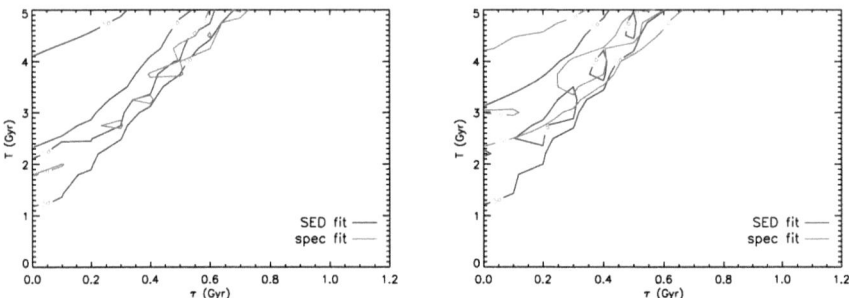

Figure 4.7: Confidence regions of the fit: 1σ and 3σ confidence regions, in the $t_{burst} = 0$ (no secondary burst) plane, of the fit to the average SED (blue) and spectrum (red) of the GOODS (left) and RDCS 1252 (right) samples.

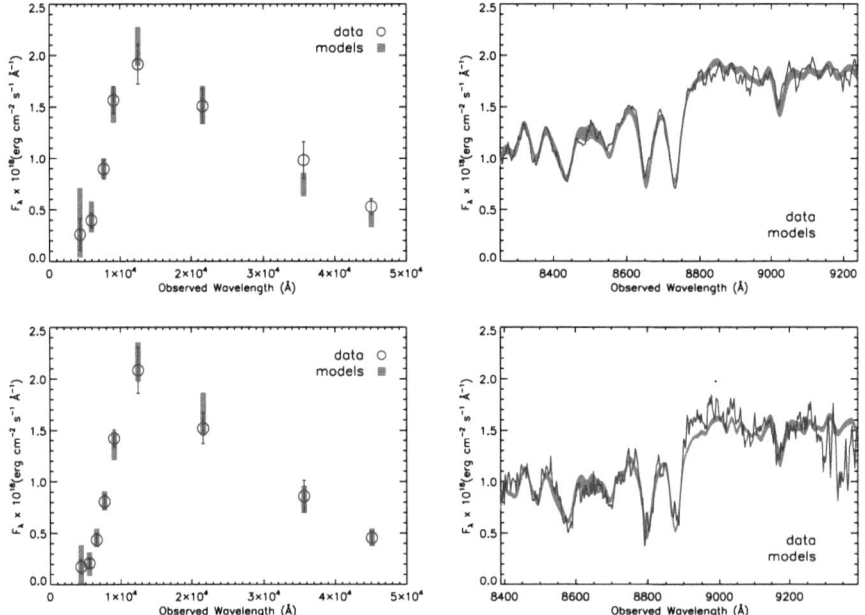

Figure 4.8: Average SED (left) and spectrum (right) of the early-type galaxies in the GOODS (up) and RDCS 1252 (down) samples, in blue, and best fitting models within the 3σ confidence regions of the fit (red).

4.3 Results

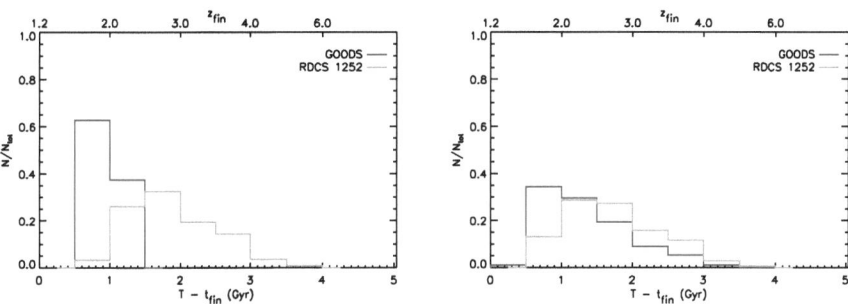

Figure 4.9: Final formation times of the GOODS and RDCS 1252 samples: fraction, weighted by χ^2, of best fitting models to the spectrophotometric (left) and photometric (right) data, within the 3σ confidence intervals, as a function of the final formation redshift z_{fin} and look-back time since $z = 1.24$, $T - t_{fin}$. The blue histogram corresponds to the GOODS field sample and the red one to the RDCS 1252 cluster sample.

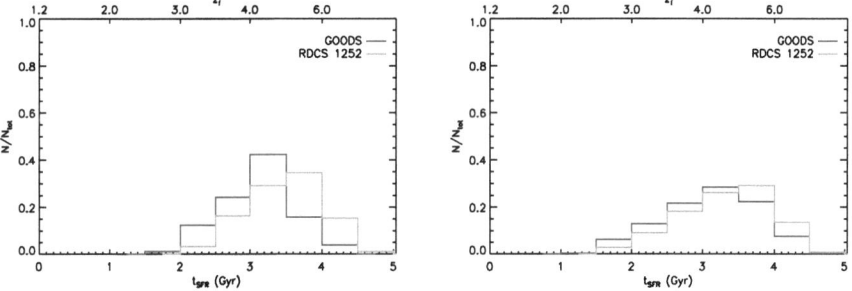

Figure 4.10: Star formation weighted ages of the GOODS and RDCS 1252 samples: χ^2-weighted fraction of best fitting models to the spectrophotometric (left) and photometric (right) data, within the 3σ confidence intervals, as a function of the star-formation weighted age t_{SFR} and corresponding formation redshift z_f. The blue histogram corresponds to the GOODS field sample and the red one to the RDCS 1252 cluster sample.

	t_{SFR} (Gyr)	z_f	$T - t_{fin}$ (Gyr)	z_{fin}
GOODS	$3.1^{+0.5}_{-0.5}$	$3.4^{+1.2}_{-0.7}$	$0.9^{+0.2}_{-0.2}$	$1.6^{+0.1}_{-0.1}$
RDCS 1252	$3.5^{+0.5}_{-0.5}$	$4.2^{+1.8}_{-1.0}$	$1.9^{+0.7}_{-0.5}$	$2.1^{+0.7}_{-0.4}$

Table 4.2: Mean star formation weighted age, formation redshift, final formation time and final formation redshift, of the models within the 3σ confidence regions of the fit to the stacked spectrophotometric data of the GOODS and RDCS 1252 samples.

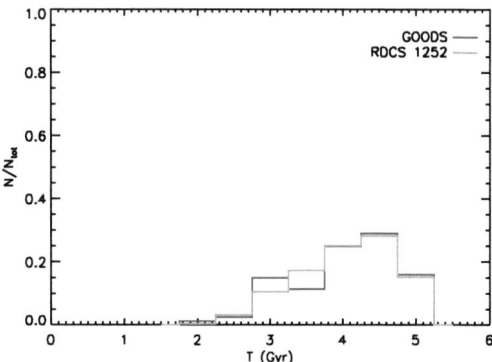

Figure 4.11: T values of the GOODS and RDCS 1252 samples: χ^2-weighted fraction of best fitting models, within the 3σ confidence regions, to the spectrophotometric data of the GOODS (blue) and RDCS 1252 (red) samples as a function of the time T since the onset of star formation.

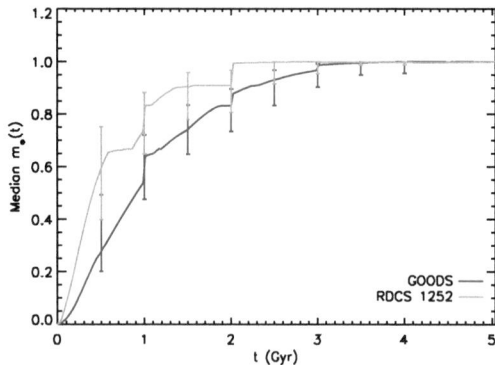

Figure 4.12: Stellar mass growth histories of the GOODS and RDCS 1252 samples: median stellar mass fraction of the best fitting models as a function of the time since the onset of star formation. The blue curve corresponds to the GOODS field sample and the red one to the RDCS 1252 cluster sample.

4.3 Results

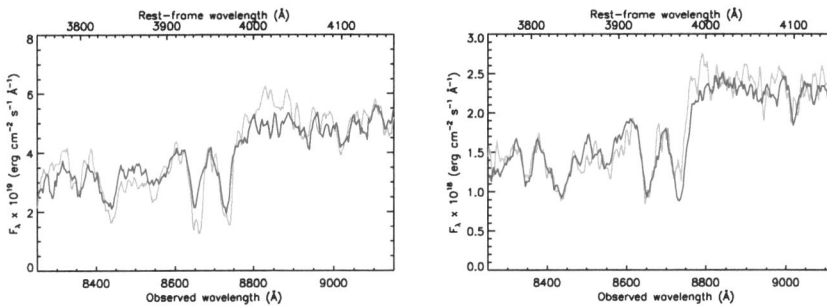

Figure 4.13: Composite spectrum of low mass ($M_\star < 1.2 \times 10^{11} M_\odot$, left) and high mass ($M_\star \geq 1.2 \times 10^{11} M_\odot$, right) galaxies in the GOODS field (blue) and RDCS 1252 cluster (red) samples. The spectra have been smoothed by 3 pixels ($\sim 7.5 \text{ Å}$).

ated in Fig. 4.13, where each sample is divided equally in two mass bins. At lower masses, the average spectrum of the cluster galaxies has a more pronounced 4000 Å break and lower star formation weighted ages than the field sample, while such a difference becomes negligible for the most massive galaxies. This elucidates how the time scale of the star formation activity in the cluster environment is shorter than in the field.

It is important to remember that the GOODS field sample is more deficient in lower mass galaxies with respect to the total photometric sample than the RDCS 1252 cluster sample, as seen in Fig. 4.5. The relative incompleteness of the GOODS sample compared to the RDCS 1252 sample reinforces the conclusions drawn from Fig. 4.9 and 4.10 as, if the field sample were corrected for completeness, the average SED and spectrum would be expected to appear bluer and thus younger, amplifying the difference between the star formation timescales of the field and the cluster. As widely reported in the literature, we also found that less massive galaxies are best fitted with younger stellar population models, the so-called "downsizing" (e.g. Cowie et al. [1996]). Unfortunately, we did not have enough statistics to study whether at this redshift the mass-age correlation varies from cluster to field.

4.3.1 Simulations

To assess the constraining power of the spectrophotometric fit when using the low signal-to-noise spectra of our samples, we performed a set of Monte-Carlo simulations using the model corresponding to the best fit parameters of the stacked spectrophotometry of each sample; for the GOODS sample we find $\{T = 4.25 \text{ Gyr}, \tau = 0.6 \text{ Gyr}, t_{burst} = 1 \text{ Gyr}, A = 0.5\}$, for RDCS 1252 $\{T = 5 \text{ Gyr}, \tau = 0.5 \text{ Gyr}, t_{burst} = 3 \text{ Gyr}, A = 0.2\}$. We carried out a hundred simulations by adding to the model SED and spectrum an amount of noise equal to

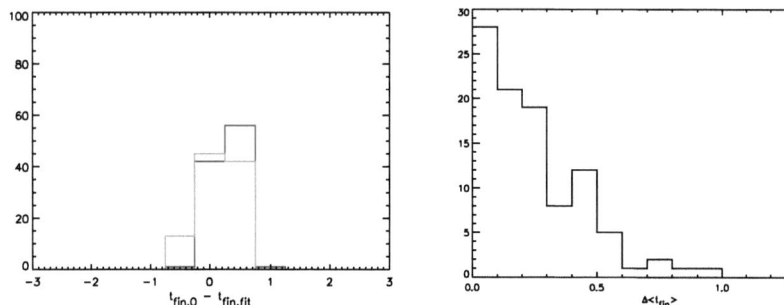

Figure 4.14: Left: distribution of the residuals of final formation times obtained by perturbing the best fit model of the GOODS (blue) and RDCS 1252 (red) samples. Right: distribution of the difference between the mean look-back time to the final formation time of 100 pairs of randomized samples.

that of the stacked spectrophotometric data and fitting the perturbed spectrophotometric data with the grid of models described in Section 4.2. As shown in Fig. 4.14 (left), we did not find any appreciable differential bias between the two samples.

To determine whether the difference between the cluster and field samples is simply due to a selection effect, we also considered pairs of randomized samples constructed from both field and cluster galaxies. We first converted the b_{435}, v_{606} and J magnitudes of the GOODS sample to B, V and J_s, as used in RDCS 1252. For this purpose, we integrated the spectra of the grid of models to derive the photometric transformations at $z \simeq 1.2$:

$$B = b_{435} - 0.021 \times (b_{435} - v_{606}) + 0.0014 \quad (4.2)$$
$$V = v_{606} + 0.32 \times (b_{435} - v_{606}) + 0.081 \quad (4.3)$$
$$J_s = J - 0.03 \times (J - K_s) + 0.013 \quad (4.4)$$

We ignored the R band data of RDCS 1252 as there is no comparable passband in the GOODS photometry. We constructed a hundred sample pairs by drawing 21 galaxies from the GOODS and RDCS 1252 samples using a uniform random distribution to constitute the "cluster" sample and assigned the rest to the "field" sample. We then fitted the synthetic spectrophotometric data of these pairs using the same grid of models described in Section 4.2. As a measure of the difference between the two pseudo-populations, we considered the absolute difference in mean look-back time to the final formation time $\Delta < T - t_{fin} >$ as it is more sensitive to differences in star formation histories than the mean star formation weighted age. Fig. 4.14 shows that the distribution of $\Delta < T - t_{fin} >$ in the mock cluster and field sample pairs is peaked at $\Delta < T - t_{fin} >= 0$ and that the value of $\sim$1 Gyr derived from the GOODS and RDCS 1252 samples lies well outside the bulk of the distribution.

4.3 Results

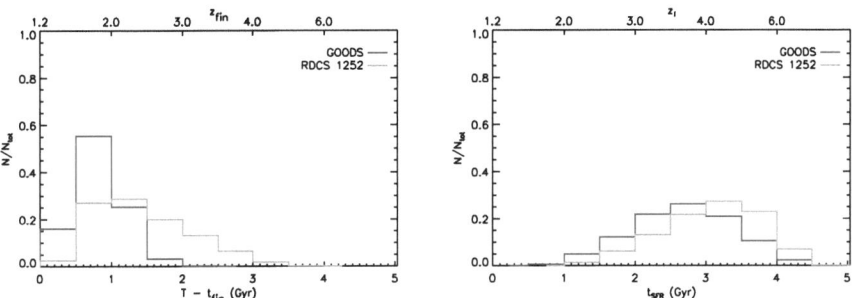

Figure 4.15: Ages of the GOODS and RDCS 1252 samples: χ^2-weighted fraction of best fitting Maraston ([2005]) models, within the 3σ confidence regions, to the stacked spectrophotometric data of the GOODS (blue) and RDCS 1252 (red) samples, as a function of the look-back time from $z = 1.24$ to the final formation time $T - t_{fin}$ (left) and star formation weighted age t_{SFR} (right).

This strongly implies that the galaxies in the field and cluster samples are drawn from two distinct populations.

4.3.2 Considerations on spectral synthesis models

In this analysis, we did not take into account any uncertainties in the model templates. However, if systematic discrepancies were present in the model spectra, in particular in some important spectral features such as the Balmer absorption lines or the 4000Å break, the χ^2 fit, and thus the best fit solutions, would be significantly affected. To test the issue of model dependency, we fitted the stacked spectrophotometric data with composite stellar population models computed from Maraston ([2005]) templates at solar metallicity, using the same initial mass function and grid of star formation histories. Fig. 4.15 show the χ^2-weighted distribution of star formation weighted age and final formation time of the best fitting models. In this case, we found a difference of 0.3 to 0.5 Gyr between the two samples. This small value is likely due to the lower resolution of the M05 spectral templates (15 Å, compared to 3 Å for the BC03 templates), as we could reproduce it by using BC03 higher-resolution templates downgraded to the resolution of the M05 templates.

In the rest-frame wavelength range of the sample spectra, the Maraston ([2005]) templates use the theoretical BaSeL (Lejeune et al. [1998]) spectral library, while the Bruzual & Charlot ([2003]) templates use the empirical STELIB (Le Borgne et al. [2003]) spectral library. Furthermore, they use different approaches when computing SSP templates, as explained in Chapter 2. Therefore, the BC03 and M05 templates are unlikely to be affected by the same systematic uncertainties. In addition, the stellar library used by the BC03 templates has a signal-to-noise ratio of ~50 per pixel, much higher than the S/N of the

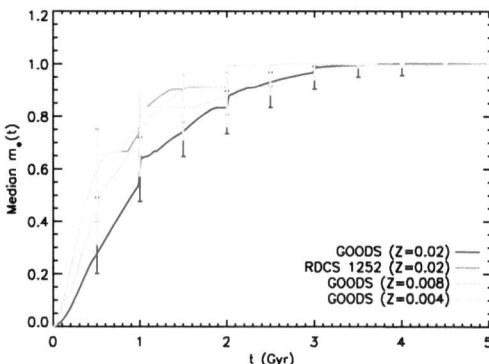

Figure 4.16: Mass growth history as a function of metallicity: median stellar mass fraction of the best fitting models as a function of the time since the onset of star formation, for RDCS 1252 at solar metallicity and GOODS at solar and sub-solar (Z=0.008, 0.004) metallicities.

stacked spectra, which means that flux uncertainties in the template spectra are negligible. We conclude that, while the star formation histories and absolute age values are somewhat model dependent, there is indeed a difference in time scales between cluster and field star formation histories.

4.3.3 Considerations about metallicity and dust

Since age and metallicity have a similar effect on the spectrum of a stellar population, the difference observed between the field and cluster samples might not actually correspond to distinct star formation histories but to a difference in metal content. To quantify this effect, we compared the averaged field galaxy data to sub-solar metallicity models computed using the Bruzual & Charlot ([2003]) templates and the same grid of parameters described in Section 4.2, keeping the cluster galaxies at solar metallicity.
As shown in Fig. 4.16, the star formation histories of the cluster and field samples coincide if the cluster is assumed to be a factor two to five more metal rich than the other. This ad hoc assumption does not appear to be realistic, however. At low redshift, the variation in metallicity of early-type galaxies with environment was found to be less than 0.1 dex (Thomas et al. [2005]), with field elliptical galaxies being more metal-rich than their cluster counterparts. As the bulk of star formation in cluster and field elliptical galaxies is understood to have happened at $z > 2$ (Renzini [2006]), it is unlikely that the relative metallicity of field and cluster ellipticals is much different at $z \sim 1.2$ than at low redshift. Metallicity differences within each sample are also unlikely to account for the observed age difference. Since the metallicity of early-type galaxies increases with mass (e.g. Tremonti

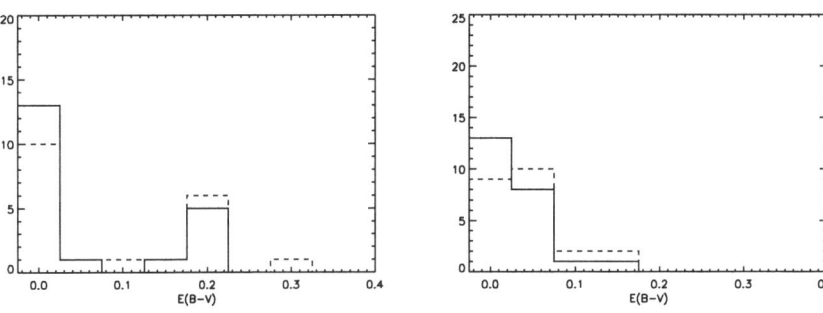

Figure 4.17: Dust in GOODS and RDCS 1252: distribution of best fit (solid histogram) and maximum, within the 3σ confidence interval (dashed histogram), E(B-V) of the fit to the SEDs of early-type galaxies in the GOODS (left) and RDCS 1252 (right) samples.

et al. [2004], Thomas et al. [2005]), a high mass galaxy will appear older than a low mass one when fitted with single metallicity models. As the GOODS sample is biased toward high masses with respect to the RDCS 1252 sample, a metallicity gradient within the samples would lead to an underestimate of the difference between the two populations.

As with metallicity, age and dust content are largely degenerate when fitting SEDs. However, since we used high-resolution spectra as well as SEDs, our fitting method is less sensitive to dust than to metallicity. Furthermore, there is no evidence that dust structures in early-type galaxies contribute significantly to their mid-IR spectra (e.g. Temi et al. [2005]) or are correlated with environment (e.g. Tran et al. [2001]). Nevertheless, we performed a fit on the SEDs of the individual galaxies using simple τ-models and assuming a Cardelli et al. ([1989]) extinction law. We considered values of the color excess $E(B-V)$ ranging from 0 to 1 with increments of 0.05 and, in the case of RDCS 1252, used only models with $z_f \geq 2.4$, as per Lidman et al. ([2004]). We found that the SEDs of the early-type galaxies in our cluster and field samples are best fitted by models with little or no dust, as shown in Fig. 4.17. This result is consistent with Rettura et al. ([2006]) and supports the dust-free assumption.

4.3.4 Rest-frame far-UV flux

Rettura et al. ([2008]) have analyzed VLT/VIMOS (LeFevre et al. [2003]) U-band data of the same sample of early-type galaxies. This corresponds, at $z \sim 1.2$, to the far-UV ($\lambda \sim 1700\text{Å}$) regime, where hot, short-living stars emit most of their light. The rest-frame far-UV can thus be used to constrain the ongoing or recent star formation in early-type galaxies of the GOODS and RDCS 1252 samples. Their deep U-band imaging reached a depth of $U = 28.27$ mag in GOODS and $U = 27.3$ mag in RDCS 1252. Most (75%) of

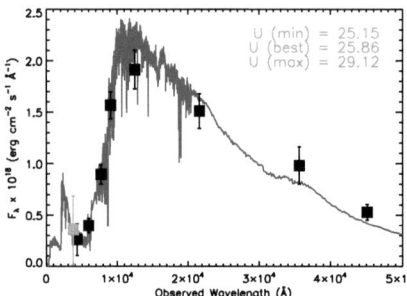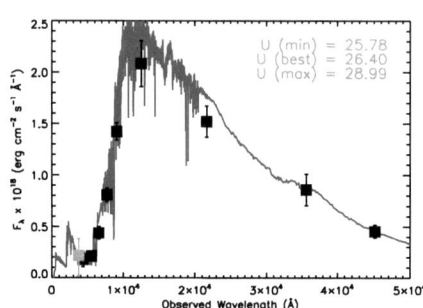

Figure 4.18: Predicted U-band fluxes of the GOODS and RDCS 1252 samples: best fit model (blue) to the average SED (black squares) of galaxies in the GOODS (left) and RDCS 1252 (right), with predicted U-band flux (red). The red error bars show the range of U-band flux values of the models within the 1σ confidence region of the fit to the stacked SED.

the field early-type galaxies were detected in the U-band images, but none of the cluster galaxies. As a result, they compared the median stacks of the field and cluster samples. After accounting for the ~ 0.4 mag larger galactic extinction in the field of RDCS 1252 with respect to GOODS, they found that field galaxies are at least 0.5 mag brighter in the U-band than cluster galaxies in the same mass range. This > 0.5 magnitude difference is consistently predicted by the best fit models to the spectrophotometric data, and is indicative of the longer formation time-scale of the field galaxies.

4.4 Comparison with semi-analytic models

High-redshift clusters such as RDCS J1252.9-2927 represent an extremely biased environment for galaxy formation where the effectiveness of galaxy evolution processes is most enhanced. Their properties thus provide a sensitive test of models of galaxy formation and evolution. In particular, the tightness of the red sequence at redshifts greater than unity (e.g. Blakeslee et al. [2003], Mei et al. [2006a], [2006b]) constitutes a strong constraint on hierarchical formation models, in which galaxy growth is a gradual process. We compared the results of our spectrophotometric analysis to the predictions of a state-of-the-art semi-analytic model of galaxy formation by Menci et al. ([2002], [2004], [2005]). Here we describe the principal characteristics of the semi-analytic model:

- The primordial dark matter density field is described is described as a random Gaussian density field with a λCDM power spectrum (Spergel et al. [2007]). The dark matter halos of the proto-galaxies collapse from overdense regions and merge at a rate described by the Extended Press & Schechter model (e.g. Bond et al. [1991]).

4.5 Scatter of the red sequence

- The description of baryonic processes includes cooling at the center of galaxies, the settling of the gas into disks, the conversion of gas into stars and stellar feedback. Starbursts triggered by fly-by or merger interactions are also included.

- The growth of supermassive black holes by cold gas inflow is described according to the model of Cavaliere & Vittorini ([2000]) and the treatment of feedback due to accretion is derived from Lapi et al. ([2005]).

- Stellar light emission is computed by convolving the star formation rate with model spectra from population synthesis models.

This model thus includes all candidate processes for the acceleration and subsequent quenching of star formation in dense environments. In Fig. 4.19, we show a comparison of the χ^2-weighted distribution of best fitting models to the GOODS and RDCS 1252 samples as a function of star formation weighted ages and the predictions of the semi-analytic model for galaxies belonging to the red sequence and with stellar masses in the same range as the GOODS and RDCS 1252 samples. Fig. 4.20 shows the model prediction for the mass growth histories of field and cluster galaxies compared to the median mass growth of the best fitting models to the stacked spectrophotometric data of the GOODS and RDCS 1252 samples. We found a good agreement overall between the model predictions for the star formation weighted ages and the ages derived from the stacked spectrophotometric data of the field and cluster samples. In the semi analytic model, the peak values of the stellar ages in the field and clusters do not differ much but the age distribution of field galaxies shows a larger scatter, towards younger ages. This is especially evident in Fig. 4.20, where the average stellar mass growths of cluster and field red sequence galaxies are not significantly different but the field shows a much larger scatter towards late mass growths. On the other hand, whereas the model predicts roughly the same average mass growth in both environments, the best fitting models to the spectrophotometric data suggest that stellar mass forms earlier in cluster galaxies, as seen in Fig. 4.12. While this discrepancy could be due to differences in sample and methodology (the model mass growth histories are computed from known star formation histories while the mass growth histories of the GOODS and RDCS 1252 samples are derived from the star formation histories of a grid of Bruzual & Charlot models), the red sequence scatters predicted by the semi analytic model are still larger than the observed scatters (see Menci et al. [2008]). This indicates the need for a faster assembly time scale in clusters in the models.

4.5 Scatter of the red sequence

Finally, we measured the synthetic scatter of the rest-frame $(U - B)_z$ color, $\Delta(U - B)_z$, of the early-type galaxies on the red sequence of RDCS 1252. We used the morphological classification and selection criterion described by Blakeslee et al. ([2003]) and selected E and S0 galaxies brighter than $z_{850} = 24.0$ whose spectroscopic or photometric redshift

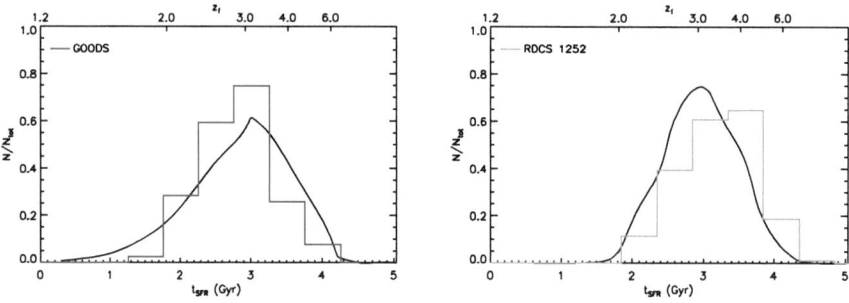

Figure 4.19: Ages of model galaxies: distribution of stellar ages (solid line) of galaxies in the mass range $5 \times 10^{10} \leq M_\star/M_\odot \leq 5 \times 10^{11}$ at $z = 1.2$ in the field (left) and in clusters (right). The red histograms represent the χ^2-weighted distribution of star formation weighed ages of the best fitting models to the stacked spectrophotometric data of the GOODS (left) and RDCS 1252 (right) samples.

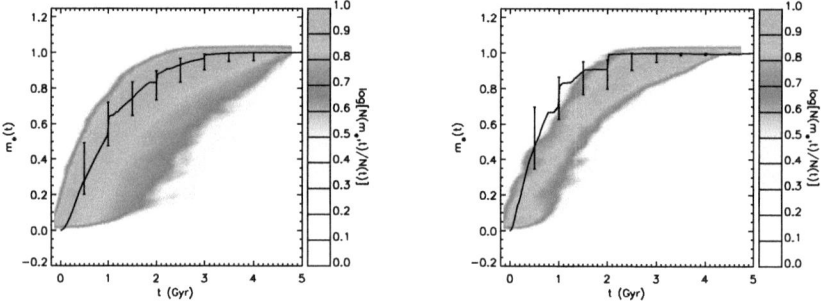

Figure 4.20: Stellar mass growth of model galaxies: stellar mass growth history of galaxies at $z = 1.2$ with stellar masses in the range $5 \times 10^{10} \leq M_\star/M_\odot \leq 5 \times 10^{11}$ for in the field (left) and cluster (right) environments. The mass fraction $M_\star/M_{\star,0}$ formed at time t is computed as the stellar mass contained in all progenitors of each galaxy divided by its final (at $z = 1.2$) stellar mass. The color code represents the number of galaxies with given $M_\star/M_{\star,0}$ at time t normalized to the total number of galaxies at that time. The data points show the median stellar mass fraction as a function of time of the best fitting models to the stacked spectrophotometric data of the GOODS (left) and RDCS 1252 (right) samples.

4.5 Scatter of the red sequence

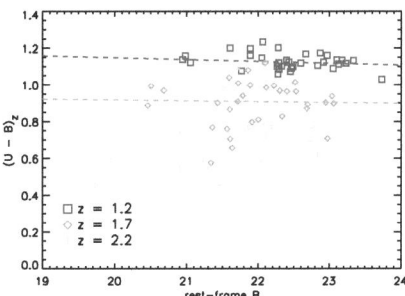

Figure 4.21: Predicted rest-frame $(U-B)_z$ color-magnitude relation of early-type galaxies in RDCS J1252.9-2927 based on the best fit models to the 9-band SEDs, at $z = 1.24$, 1.7 and 2.5. The dashed lines show the best fit linear correlations.

matched the redshift of the cluster. This criterion yielded 33 red sequence galaxies. Rest-frame $(U-B)_z$ colors were obtained for each galaxy from the model that best fitted the 9-band SED. For this purpose, we used only simple τ-models. The best fit models were then devolved from $z = 1.237$ to $z = 2.2$. At each redshift, the scatter was measured by applying a 3σ clipping around the linear best fit, as customary in the literature, and model galaxies within 3σ of the linear best fit were considered as being part of the red sequence. A few additional galaxies drop out of the red sequence sample before $z = 2.2$ as the redshift becomes larger than $z(T_0)$, where T_0 is the age of the best fit model. We estimated the uncertainty on the scatter at each redshift by randomly perturbing the SED of each galaxy, assuming Gaussian errors on the fluxes. Fig.4.21 shows the predicted color-magnitude relation of RDCS 1252 at three different redshifts.

At $z = 1.237$ we found a good agreement with the 0.05 intrinsic scatter value measured by Blakeslee et al. ([2003]). Since the best fit models used here to estimate the red sequence scatter were derived from the full 9-band SED, the synthetic colors are naturally less sensitive to uncertainties on the individual bands, especially as the U and B rest-frame passbands lie in the middle of the wavelength range covered by the SED. Fig. 4.22 shows that the red sequence dissipates at $z \sim 1.9$, reaching scatter values comparable with those observed in the red population of the field up to $z \sim 2$ (Cassata et al. [2008]). At $z \sim 1.9$, 25 galaxies out of the original 33 are still included in the red sequence sample. This redshift is also consistent with the median final formation redshift derived from the analysis of the stacked spectrophotometric data. A similar analysis using $(U-V)_z$ and $(U-R)_z$ colors yielded the same result. We conclude that the red sequence in such a massive cluster is established in ~ 1 Gyr and it is therefore not surprising that recent analysis of forming red sequences in protoclusters at $z > 2$ find a significant scatter (Kodama et al. [2007], Tanaka et al. [2007], Zirm et al. [2007]).

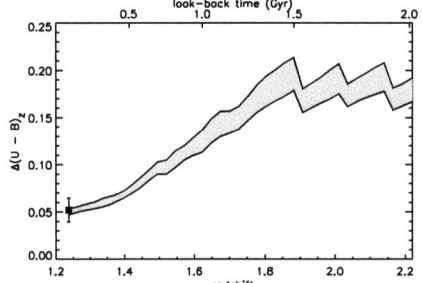
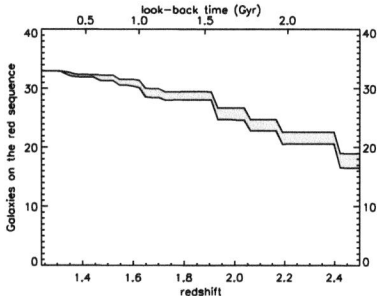

Figure 4.22: Left: predicted rest-frame $(U-B)_z$ scatter of the red sequence of RDCS J1252.9-2927 extrapolated to $z = 2.2$. The shaded area represents the rms dispersion of the models that best fit the SEDs of the individual red sequence galaxies. The filled square is derived from the intrinsic $i_{775} - z_{850}$ scatter measured by Blakeslee et al. ([2003]). Right: predicted number of galaxies on the red sequence of RDCS J1252.9-2927, extrapolated to $z = 2.2$.

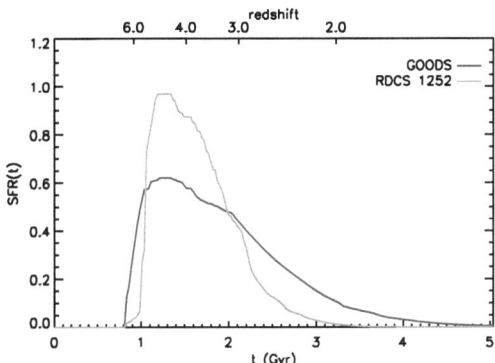

Figure 4.23: Median star formation history of best fitting models to the stacked spectrophotometric data of the GOODS field (blue) and RDCS 1252 cluster (red) samples, smoothed to 0.25 Gyr. The lower x-axis shows the cosmic time t and the upper x-axis the corresponding redshift.

4.6 Summary

We have used two homogeneous multi-wavelength datasets of a massive cluster at $z = 1.24$ and the GOODS field. The depth and number of passbands make these data arguably the best available to date at this redshift. By modeling the spectrophotometric properties, we found a small although significant difference in the star formation histories of the cluster and field populations, suggesting that the cluster galaxies form the bulk of their stars ~0.5 Gyr earlier than their counterparts in the field, with massive early-type galaxies having already finished forming stars at $z > 1.5$ in both environments. Star formation in both environments starts at the same time but the time-scale is longer in field early-types, as shown in Fig. 4.23.

While the differential analysis of the stellar population parameters of cluster and field galaxies in the same mass range convincingly shows distinct star formation histories, the absolute age difference remains model dependent. We verified that such a difference in derived star formation histories in the two environments cannot be ascribed to incompleteness of the mass selected samples, which would tend to rather increase such an effect. We also used extensive Monte-Carlo simulations to identify possible biases in the model fit and discussed the effect of inherent degeneracies such as metallicity and dust. We note that independent studies of massive early-type galaxies based on the measurement of the mass-to-light ratios of massive early-type galaxies in high- and low-density environments (Treu et al. [2005], van der Wel et al. [2005], van Dokkum & van der Marel [2007]) have reached similar conclusions. This age difference is however significantly smaller than the one ($\gtrsim 1.5$ Gyr) deduced from low redshift observations (e.g. Thomas et al. [2005], [2006],

Clemens et al. [2006]), but can be reproduced in the star formation weighted ages if we assume that the average population of field early-type galaxies assembled $\sim 10\%$ of its stellar mass at $z < 1$. We note that this is more likely to be an upper limit, as the conclusions of the low redshift studies are based on luminosity-weighted ages, i.e. ages derived from SSP templates, which are more sensitive to the last episode of star formation.

We also used the best fit star formation histories from the 9-band SEDs of the red sequence galaxies in RDCS 1252 to predict that a tight ($\Delta(U-B)_z = 0.05$ mag) red sequence at $z \sim 1.2$ is established over approximately 1 Gyr and dissolves by $z \approx 1.9$. This implies that for massive clusters, which have long reached virialization by redshift 1.2, a $(U-B)_z$ color scatter well above 0.1 mag is expected (i.e. no significant red sequence).

These results suggest that, in order to observe the actual build-up of the red sequence in clusters, it will be crucial to discover and study clusters in the so-called "cluster desert" at $1.5 \lesssim z \lesssim 2$. The combination of wide area surveys in the near-IR (e.g. with the VISTA telescope) and IR (with *Spitzer*/IRAC, e.g. Eisenhardt et al. [2004]) surveys, with X-ray and Sunyaev-Zel'dovich searches (e.g. Staniszewski et al. [2008]) will enable these studies in the near future.

Chapter 5

Star formation histories in a dense environment at $z \sim 0.84$

As stated in the previous Chapters, the study of the stellar population properties of galaxies across a wide range of environments is crucial to understanding the processes that drive galaxy formation and evolution. At high redshift, this is typically done by comparing cluster galaxies to the field population. Some of the problems that can affect cluster to field comparisons are cosmic variance, relative incompleteness (see Chapter 4) and inhomogeneous datasets. On the other hand, galaxy clusters alone provide a range of environmental densities, from the near-field outskirts to the extremely biased core, in which to investigate galaxy evolution with all the advantages of homogeneous data. At high redshifts however, the long integration times needed to obtain a usable signal to noise ratio often means that only the most luminous cluster galaxies can be observed, resulting in low number statistics and rendering more difficult the interpretation of results. With the massive cluster RX J0152.7-1357 at $z = 0.837$, we have for arguably the first time the combination of high redshift and large dataset necessary to investigate the modes of galaxy evolution in a single high density environment at more than half the Hubble time. RX J0152.7-1357 was found independently by at least two X-ray surveys (Ebeling et al. [2000]) and exhibits a complex structure, with two main components separated by $\sim$1.6' which appear to be in the early stages of merging. The wealth and quality of the spectroscopic and photometric data available for this cluster (more than 130 confirmed members with 5-band photometry and optical spectroscopy) allowed us to carry out a detailed study of the variation of stellar population parameters along the red sequence, giving us clues on the mode and time scales with which galaxies migrate from the blue to the red population in high density environments.

This Chapter is organized as follows. In Section 5.1, we describe our spectroscopic sample of 134 galaxy members of RX J0152.7-1357, the selection of early-type galaxies and their grouping into luminosity, color, mass and environmental density bins. In Section 5.2, we present the results of our analysis on the composite spectrophotometric data of each bin. In Section 5.3, we discuss possible biases due to metallicity effects and internal dust.

5.1 Observations and sample selection

RX J0152.7-1357 was observed in the F625W (hereafter r_{625}), i_{775} and z_{850} bands with ACS on the HST (Blakeslee et al. [2006]) as well as in the J and K_s bands with SofI (Moorwood et al. [1998]) on the ESO NTT (Demarco et al. [2005]). Magnitudes were measured using SExtractor (Bertin & Arnouts [1996]) in apertures of 0.75" for the ACS bands and 1" for the SofI bands, corrected to apertures of 2" and 5" respectively, which should enclose all the light of the galaxies in these passbands. Extinction corrections were calculated from Schlegel et al. ([1998]) using the NED galactic extinction calculator[1] (see Chapter 4). The extinction corrections are 0.04 mag in r_{625}, 0.03 in i_{775}, 0.023 in z_{850}, 0.014 in J and 0.005 in K. For this study, we also used spectra of 102 cluster members confirmed by Demarco et al. ([2005]), complemented with spectra of 32 newly confirmed cluster members obtained using FORS2 on the VLT, with a resolution of about 13Å. At $z = 0.837$, the H_δ absorption feature is blended with an atmospheric A-band line. Because of the possibility of sky over- or undersubtraction, we could not use the H_δ line itself to estimate the signal-to-noise ratio of our spectra. Instead, the S/N was based on the ratio between the mean and r.m.s flux in two pseudocontinuum windows straddling the H_δ absorption line, namely 4041.6 to 4079.75 Å and 4128.5 to 4161.0 Å (Worthey & Ottaviani [1997]). As those two windows do not represent the true continuum, the signal-to-noise ratios of our galaxy spectra are likely slightly underestimated.

As stated above, the richness and quality of the spectrophotometric data available for this cluster, notably the 91 confirmed early-type members, makes it possible to study the stellar population properties of cluster galaxies as a function of their intrinsic properties and environment. We performed a detailed analysis of the early-type galaxy population of RX J0152.7-1357 as a function of magnitude, color, mass and environment. For this latter part, we grouped the early-type galaxies according to their angular position and the local dark matter density, using the detailed dark matter density map of the cluster by Jee et al. ([2005]). Table 5.1 summarizes the definitions of the different bins and regions.

5.1.1 Galaxy colors and luminosities

To understand how galaxies evolve into and on the red sequence, the red galaxy population was divided into 9 bins. The cluster members were separated into active and passive galaxies using the following criterion:

- A color cut at $r_{625} - K_s = 2.3$ was arbitrarily set to distinguish the red galaxy population from the blue. For simplicity, this was defined as a line of constant color, but we note that a cut parallel to the red sequence would not have yielded a different sample. As shown in Fig. 5.3, this neatly separates the red sequence of RX J0152.7-1357 from the "blue cloud" of actively star forming galaxies. The r_{625} and K_s filters appropriately straddle the 4000 Å break, as shown in Fig. 5.2, and the K_s band is, at $z \sim 0.84$, unaffected by recent star formation (Stanford et al. [1998]).

---
[1] http://nedwww.ipac.caltech.edu/

Figure 5.1: 2.5'×2.5' (∼1.1 Mpc at $z = 0.84$) region centered on RX J0152.7-1357, from a color composite image of HST/ACS r_{625}, i_{775} and z_{850} (Blakeslee et al. [2006]). The north and south substructures (see 5.1.4) are visible on the top and bottom of the image respectively.

Bin	N_{stack}	Bin definition	Comments
1	25	$18.5 < K_s < 20.3$ and $2.3 < r_{625} - K_s < 4.5$	Bright red sequence
2	28	$19.8 < K_s < 21.1$ and $2.3 < r_{625} - K_s < 4.5$	Middle red sequence
3	26	$20.6 < K_s < 23.0$ and $2.3 < r_{625} - K_s < 4.5$	Faint red sequence
4	4	$20.3 < K_s < 21.9$ and $0.0 < r_{625} - K_s < 2.3$	Bright blue cloud
5	2	$21.9 < K_s < 23.5$ and $0.0 < r_{625} - K_s < 2.3$	Faint blue cloud
6	10	$18.5 < K_s < 20.75$ and $2.3 < r_{625} - K_s < RSfit - 0.1$	Bright, blue red sequence
7	14	$18.5 < K_s < 20.75$ and $RSfit - 0.1 < r_{625} - K_s < RSfit + 0.1$	Bright, middle red sequence
8	9	$18.5 < K_s < 20.75$ and $r_{625} - K_s \geq RSfit + 0.1$	Bright, red red sequence
9	8	$20.75 < K_s < 23.0$ and $2.3 < r_{625} - K_s < RSfit - 0.1$	Faint, blue red sequence
10	9	$20.75 < K_s < 23.0$ and $RSfit - 0.1 < r_{625} - K_s < RSfit + 0.1$	Faint, middle red sequence
11	8	$20.75 < K_s < 23.0$ and $r_{625} - K_s \geq RSfit + 0.1$	Faint, red red sequence
12	23	$8.4 \times 10^{10} M_\odot < M \leq 3.9 \times 10^{11} M_\odot$ and $2.3 < r_{625} - K_s < 4.5$	High mass red sequence
13	18	$2.7 \times 10^{10} M_\odot < M \leq 8.4 \times 10^{10} M_\odot$ and $2.3 < r_{625} - K_s < 4.5$	Intermediate mass red sequence
14	16	$4.8 \times 10^{9} M_\odot < M \leq 2.7 \times 10^{10} M_\odot$ and $2.3 < r_{625} - K_s < 4.5$	Low mass red sequence
15	18	$\Sigma_{DM} \geq 20 \times \sigma_{DM}$	High local density
16	25	$5 \times \sigma_{DM} < \Sigma_{DM} < 20 \times \sigma_{DM}$	Intermediate local density
17	19	$\Sigma_{DM} \leq 5 \times \sigma_{DM}$	Low local density
18	13	$R \leq 1'$, North	North central sector
19	13	$1' < R \leq 2'$, North	North first sector
20	8	$2' < R \leq 3'$, North	North second sector
21	20	$R \leq 1'$, South	South central sector
22	7	$1' < R \leq 2'$, South	South first sector
23	1	$2' < R \leq 3'$, South	South second sector

Table 5.1: Bins in color-magnitude space, stellar mass, local dark matter density and distance from the cluster center. The first column gives the number of each bin and the second column indicates the number of galaxy spectra stacked within the given bin. The third and fourth columns give the definition and description of each bin respectively.

5.1 Observations and sample selection

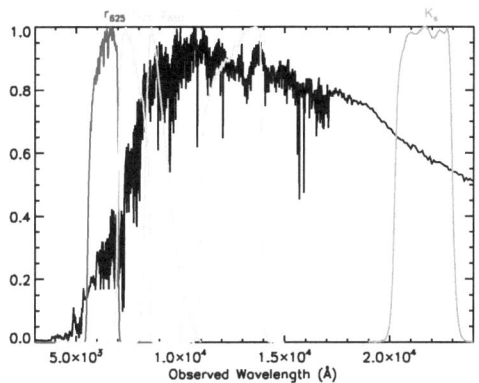

Figure 5.2: Passbands of the r_{625}, i_{775}, z_{850} (HST/ACS) and J, K_s (NTT/SofI) used in this analysis. As a reference, the model spectrum of a simple stellar population of 6.5 Gyr at solar metallicity, redshifted to $z = 0.837$, is shown. The r_{625} and K_s filters appropriately straddle the 4000 Å break and were therefore chosen for the color-magnitude diagram.

- Only red galaxies with no detectable [OII] emission were selected, using the EW([OII])> -5Å criterion of Dressler et al. ([1999]).

- Red passive galaxies were then grouped into bright, middle and faint magnitude bins, numbered (1), (2) and (3), and blue star-forming galaxies into a bright and a faint bin, (4) and (5), as shown in Fig. 5.4.

- To better explore the variations of the stellar population properties of red sequence galaxies as a function of color and magnitude, the red sequence was also divided into a "bright" region and a "faint" one. The linear fit to the red sequence,

$$r_{625} - K_s = -0.22 \times K_s + 7.75 \tag{5.1}$$

was then used to subdivide each region into blue, central and red bins ((6), (7) and (8) for the "bright" region and (9), (10) and (11) for the "faint" one), as shown in Fig. 5.4. To have a reasonable number of galaxies in each bin, the width of the two central bins was arbitrarily set to 0.2 mag in $r_{625} - K_s$.

5.1.2 Stellar mass

In a second approach, red sequence galaxies were grouped by stellar mass into three bins, (12), (13) and (14), adjusted in order to have roughly the same number of galaxies per bin. The high, intermediate and low mass ranges are 8.4×10^{10} to 3.9×10^{11}, 2.7 to 8.4×10^{10}

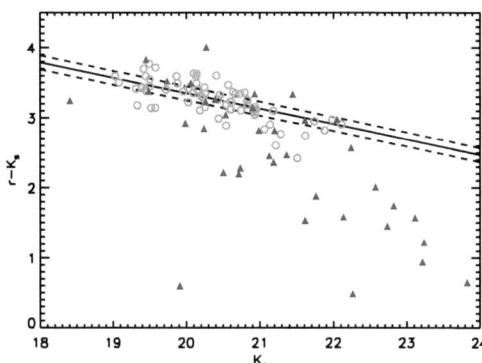

Figure 5.3: Color-magnitude diagram of the spectroscopic members of RX J0152.7-1357. The open circles indicate passive galaxies while the filled triangles show the star forming ones.

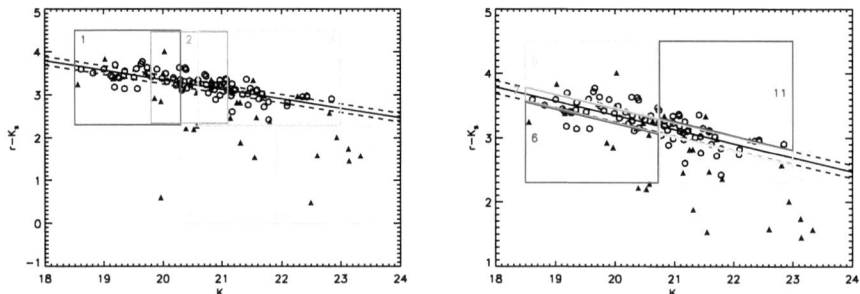

Figure 5.4: Definitions of bins in color-magnitude space. A color cut was made at $r_{625} - K_s = 2.3$ to separate blue from red galaxies.

5.1 Observations and sample selection

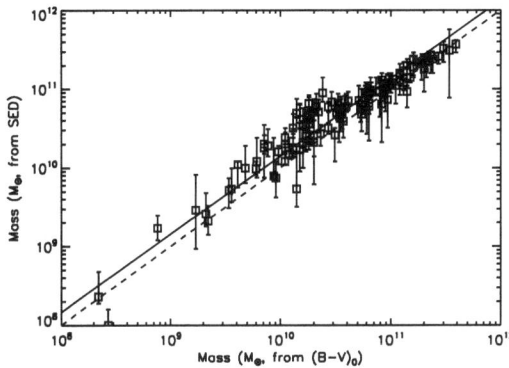

Figure 5.5: Comparison of photometric stellar masses derived from SED fitting with the masses obtained from the rest-frame $B - V$ colors. The two stellar mass estimates are consistent.

and 4.8×10^9 to 2.7×10^{10} solar masses respectively. Stellar mass estimates for the cluster members were already available from Holden et al. ([2007]), based on the mass-to-light ratio and rest-frame $B - V$ color linear relation of Bell et al. ([2003]) and consistent with estimates from dynamical measurements. However, for this analysis, we decided to recompute the stellar masses by fitting τ-models to the SEDs of the cluster galaxies, as described in Chapter 3. The templates and parameter grid are described below in Section 5.2).

As shown in Fig. 5.5, we found that the photometric stellar masses derived from SED fitting are consistent with those of Holden et al. ([2007]). The linear best fit relation is

$$M^\star_{SED} = (1.7 \pm 0.27) \times M^{0.99 \pm 0.3}_{(B-V)0} \tag{5.2}$$

The slight overestimate of the SED-derived masses with respect to those of Holden et al. ([2007]) might be due to the choice of the initial mass function, as using a higher lower mass cut-off or a top-heavy IMF like the one of Chabrier ([2003]) or Kroupa ([2001]) would result in lower mass estimates. Likewise, if the metallicity of cluster galaxies is greater than solar, their SEDs would appear older when compared to a solar metallicity model. As a consequence, the fit would tend to overestimate the near-IR fluxes and thus the stellar masses. Fig. 5.6 shows the distribution of the photometric stellar mass of galaxies in the red sequence bins.

5.1.3 Local dark matter density

Jee et al. ([2005]) performed a detailed weak lensing analysis of RX J0152.7-1357. They measured the shear signal of the cluster, using the available r_{625}, i_{775} and z_{850} ACS data

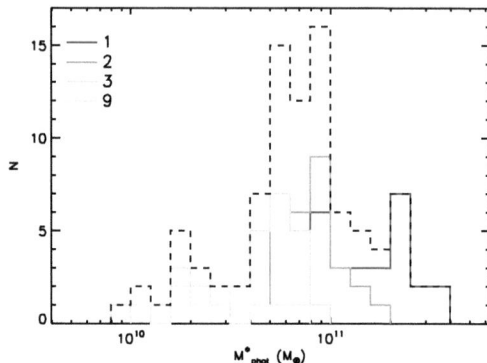

Figure 5.6: Distribution of photometric stellar masses, derived from SED fitting, of the galaxies in the red sequence bins. The dashed histogram shows the total distribution of galaxy masses.

together with the redshifts of Demarco et al. ([2005]), and reconstructed its dimensionless mass density κ. Based on the κ map of Jee et al., we grouped the red sequence galaxies into 3 cluster regions of decreasing projected mass density: $> 20 \times \sigma_{DM}$, $5 - 20 \times \sigma_{DM}$ and $< 5 \times \sigma_{DM}$ respectively, where $\sigma_{DM} = 0.0057 \times \Sigma_c$ and $\Sigma_c \sim 3650\ M_\odot pc^{-2}$ is the critical mass density of the cluster (Blakeslee et al. [2006]). These regions, named (15), (16) and (17), are shown in Fig. 5.7. Because the shear is invariant under transformations of the kind $\kappa \to \kappa' = \lambda\kappa + (1 - \lambda)$ where λ is a constant (e.g. Bradač et al. [2004]), the so-called "mass-sheet degeneracy", the projected total mass densities can only be used in a relative sense.

5.1.4 Projected angular distribution

To study the effects of environment on the stellar population properties of the cluster members, we also grouped the red sequence galaxies of RX J0152.7-1357 based on their angular distribution on the sky. Due to the complex structure of the cluster, composed of two central clumps in the process of merging, the cluster field was divided in two halves at its fiducial center (R.A. = 01h 52m 41.80s, DEC= -13° 57' 52.5") and two groups of concentric semi-annuli centered on each clump were defined, as shown in Fig. 5.8. Each semi-annulus is 1' wide. The regions associated with the northern clump ((18), (19), (20)) were labeled "North" while those associated with the southern clump ((21), (22) and (23)) were labeled "South".

5.1 Observations and sample selection

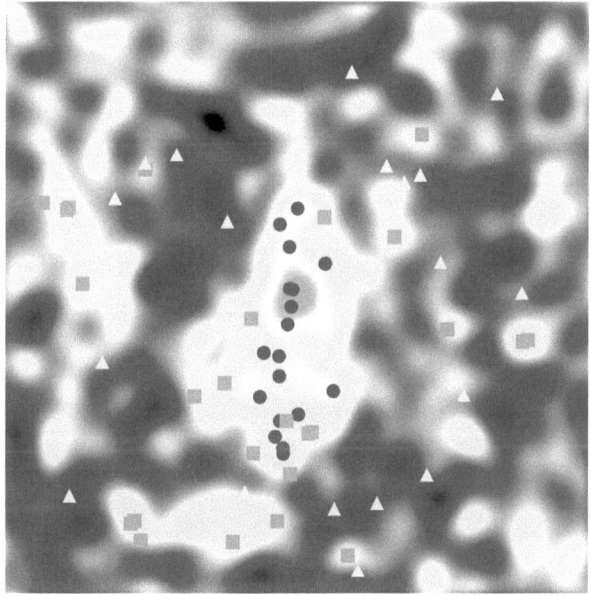

Figure 5.7: Dark matter surface density (κ) map of RX J0152.7-1357 (Jee et al. [2005]). The lowest density regions are colored blue while the highest density peaks are colored red. Passive cluster members in the highest density region (15) are indicated as blue circles, those in the intermediate density region (16) as red squares and the early-type galaxies in the lowest density region (17) as orange triangles.

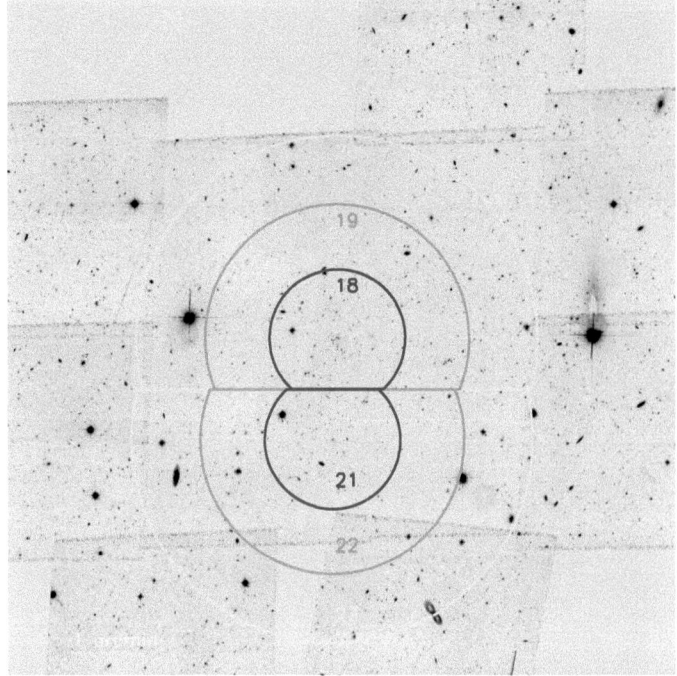

Figure 5.8: Regions used to group galaxies according to their projected angular distance. Each set of concentric semi-annuli is centered on one of the cluster clumps.

5.1 Observations and sample selection

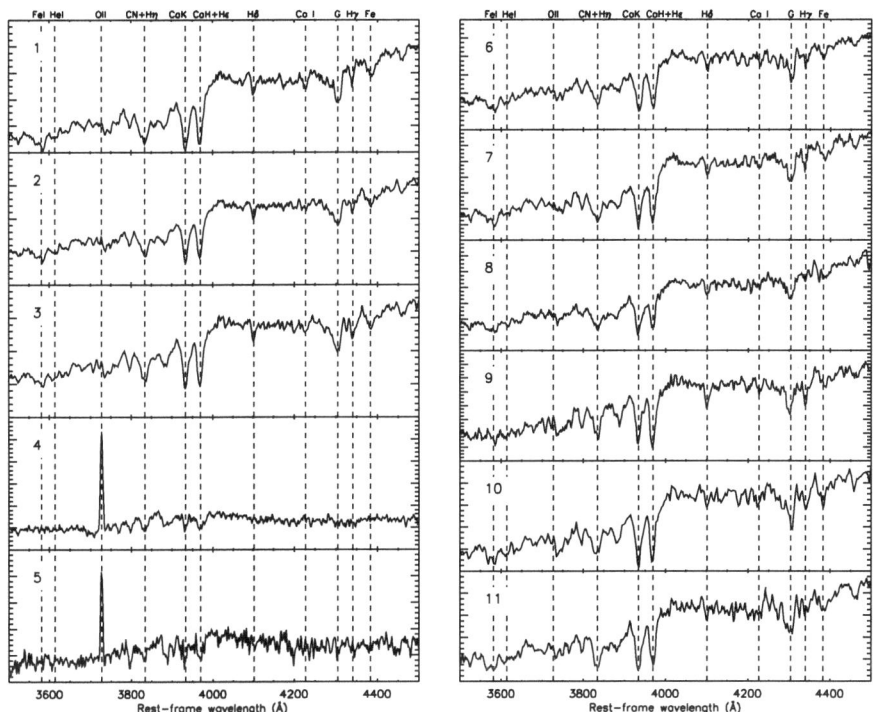

Figure 5.9: Composite spectra of red sequence galaxies in the color-magnitude diagram (left) and red sequence (right) bins. Some of the relevant spectral features are indicated by dashed lines.

5.1.5 Composite spectrophotometry

The spectra of the 134 cluster galaxies have signal-to-noise ratios ranging from 1 to 33, with a mean value of 7.6. Since the quality of most of the individual spectra is not high enough to constrain the star formation histories (see Chapter 2), the spectra were stacked on a pixel-by-pixel basis with a 1-σ rejection criterion and without iteration, as in Chapter 4. Additionally, the individual spectra were weighted by their signal-to-noise ratio and only spectra with S/N>3 were considered. Fig. 5.9 and 5.10 show the stacked galaxy spectra of each bin or region. As in Chapter 4, the total response of the detector was then normalized to the S/N value of the stacked spectrum to obtain a noise estimate as a function of wavelength. The SEDs of the individual galaxies were averaged as well.

88   5. Star formation histories in a dense environment at $z \sim 0.84$

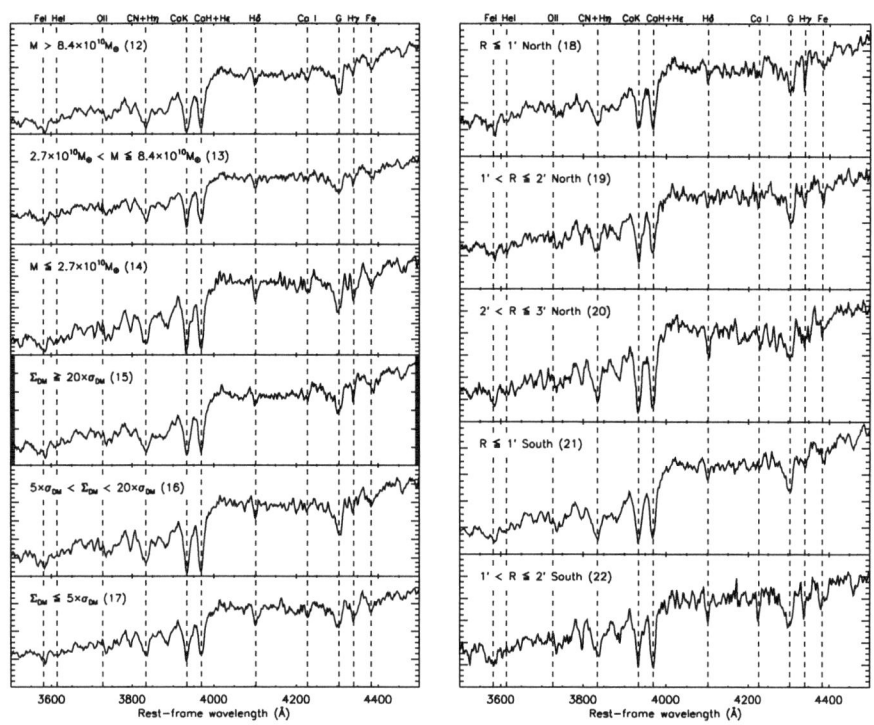

Figure 5.10: Composite spectra of red sequence galaxies selected by mass (left), local dark matter density (left) and projected angular distribution (right). Some of the relevant spectral features are indicated by dashed lines.

5.2 Stellar population modeling

As in Chapter 4, we compared the average SED and spectrum of each bin or region to a set of composite stellar population models computed using Bruzual & Charlot ([2003]) templates at solar metallicity and using the same delayed, exponentially declining star formation history described in Chapter 2. We considered values of T from 200 Myr to the age of the Universe at $z = 0.84$ with increments of $\sim$250 Myr and values of the characteristic time-scale τ from 0 to 2 Gyr with increments of 50 Myr. We assumed a Salpeter ([1955]) initial mass function with the same mass cut-offs at 0.1 and 100 $M_\odot$. This grid of τ-models was compared to the composite SED and spectrum independently using the χ^2 goodness-of-fit test described in Chapter 2. The best fitting models within the intersection of the 3σ confidence regions of the fits to the average SED and spectrum were retained. Table 5.2 summarizes the results of the spectrophotometric analysis, giving the average star formation weighted age, corresponding redshift, final formation look-back time from $z = 0.837$ and final formation redshift of each bin or region.

5.2.1 Star formation histories as a function of luminosity and color

In Fig. 5.11, we show the star formation weighted age t_{SFR} and final formation look-back time $T - t_{fin}$, from $z = 0.837$ of the best fitting models to the stacked spectrophotometric data of the red sequence bins (1) to (3) and (6) to (11). We also tested the spectrophotometric data of the blue galaxy bins (4) and (5) and provide the results of the fit in Table 5.2 for comparison. These latter values are almost certainly overestimated, however, as we ignored the [OII]λ3727 emission feature and didn't include dust extinction in the fit. The first would constrain the star formation rate at $z = 0.837$, and thus the range of star formation histories, while the second would allow for younger models to fit the observed SED equally well by making them redder. The constraints from the [OII]λ3727 emission and dust would therefore produce lower ages than those derived from dustless, passive models. We found a clear correlation between age and color and between age and luminosity, the redder and brighter bins being older. Specifically, we found that $t_{SFR} \propto K_s$ and $t_{fin} \propto 1.5(r - K_s)$, with the galaxies in the bright tip of the red sequence having formed the bulk of their stars at $z \gtrsim 3$ and finished their star formation at $z \sim 2$ while the faint end of the red sequence formed at $z \sim 2$. In particular, while bins (10) and (11) have similar star formation weighted ages and final formation times, the faint blue end of the red sequence, represented by bin (9), is very young, with a final formation redshift of $z \sim 1$, less than 1 Gyr from the cluster's epoch. Interestingly, a similar spread toward younger ages in faint, low-mass galaxies was found by Gallazzi et al. ([2006]) using a sample of early-type galaxies in the SDSS.

To further characterize the red sequence bins, we computed three different spectral indices from the composite spectra, namely the $D_n(4000)$ continuum index as defined by Balogh

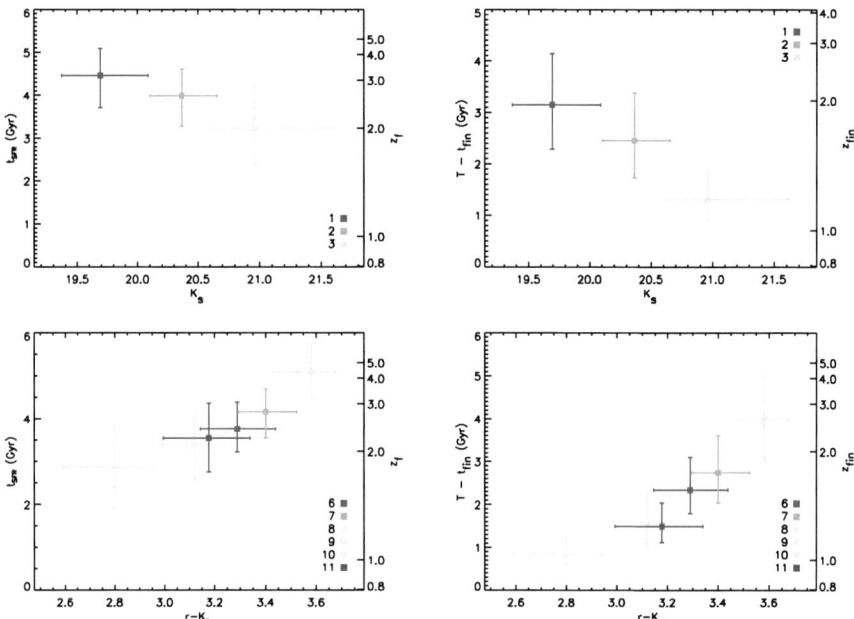

Figure 5.11: Ages of red sequence galaxies as a function of luminosity and color: mean and spread of the χ^2-weighted distributions of star formation weighted ages t_{SFR} (left) and final formation look-back times $T - t_{fin}$ (right) of best fitting models to the composite spectrophotometric data of the luminosity selected red sequence bins (1) to (3) (top) and the color selected red sequence bins (6) to (11) (bottom).

5.2 Stellar population modeling

et al. ([1999]), the Hδ_A line index (Worthey & Ottaviani [1997]) and the equivalent width of H6, a higher order Balmer feature not contaminated by metal lines (e.g. van Dokkum & Stanford [2003]) and not affected by any atmospheric line at this redshift, unlike Hδ. Although sensitive to metallicity effects, the $D_n(4000)$ index is a good indicator of the mean age of a stellar population (e.g. Poggianti & Barbaro [1997], Kauffmann et al. [2003]) while the Blamer indices trace recent star formation. Fig. 5.13 shows the comparison of the $D_n(4000)$ index with the equivalent width of the Balmer features Hδ and H6. We found that the central and red bins of the faint red sequence part (represented by bins (10) and (11)) as well as the bright part of the red sequence (represented by bins (6) to (8)) show little to no Hδ absorption with a pronounced 4000Å break, consistent with an old > 2 Gyr old population (Poggianti & Barbaro [1997]). However, the composite spectrum of the faint blue red sequence bin, (9), displays moderate Hδ absorption (EW(Hδ)> 3), classifying it as a k+a type following the Dressler et al. ([1999]) scheme, as it resembles that of a population of K stars with a significant contribution from A stars. The same effect can be seen when comparing the $D_n(4000)$ index and the equivalent width of H6. The composite spectrum of the faint blue red sequence bin is thus consistent with that of a now quiescent stellar population which has experienced a significant burst of star formation in the last 1.5 Gyr, i.e. that of a post-starburst/post-star-forming galaxy (Couch & Sharples [1987], Poggianti et al. [1999]). This suggests that, while the luminous red sequence of the cluster appears fully assembled at $z = 0.837$, the faint end of the red sequence is still in the process of being populated via the migration of low-mass ($\lesssim 5 \times 10^{10} M_\odot$) galaxies from the blue cloud as their star formation shuts off. At the cluster's redshift, this limit is broadly consistent with the critical mass found for field galaxies using the luminosity functions of both early-type and star forming galaxies (e.g. Cimatti et al. [2006], Bundy et al. [2006]). In Fig. 5.12, we show the spatial distribution of galaxies in the red sequence bins (1) and (9), as well as the star forming galaxies in the blue cloud bins (4) and (5), compared to the soft X-ray emission of the hot intracluster gas measured by Maughan et al. ([2003]) with the Chandra space telescope. The post-starburst/post-star-forming population of the faint blue red sequence bin tends to avoid the center of either clump but is distributed mostly in the 1' to 2' semi-annuli. An exception are the two galaxies of bin (9) which lie near the center of the southern clump. However, these have slightly lower redshifts than the massive central galaxies, so their apparent position in the high density region might be an effect of projection. This distribution is similar to that of k+a galaxies reported in low and intermediate redshift clusters (e.g. Dressler et al. [1999]) but different from that in some nearby clusters (e.g. Poggianti et al. [2004], Ferrari et al. [2005]). Furthermore, the spatial distribution of the galaxies of the faint blue red sequence bin does not appear to be correlated with any X-ray substructure, in contrast, for example, with the k+a population of Coma as reported by Poggianti et al. ([2004]). This suggests that the truncation of their star formation was not due to interaction with the hot cluster environment.

Interestingly, the bright red tip of the red sequence, represented by bin (8), shows the second highest Hδ absorption among the red sequence bins. This might be due to undersubtraction of the atmospheric A-band (see above). On the other hand, star-forming systems can be dust-enshrouded (e.g. Coia et al. [2005], Wolf et al. [2005]) and our selection criterion

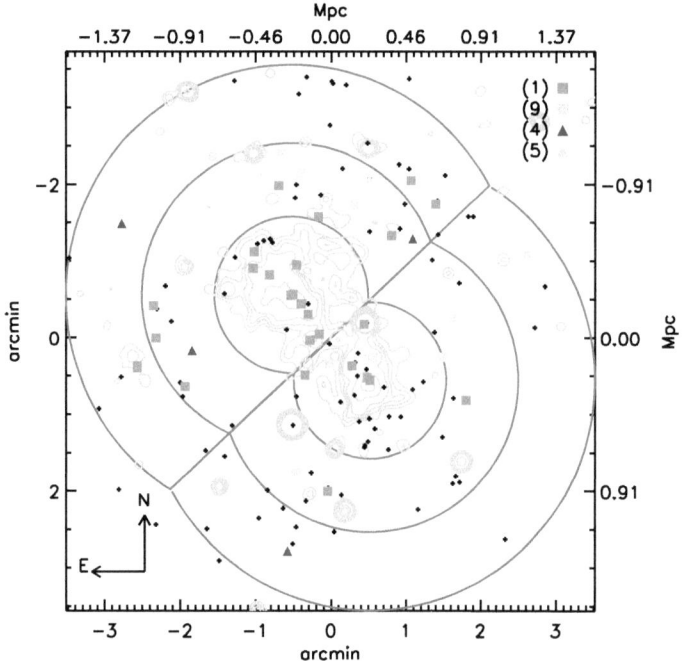

Figure 5.12: Positions on the sky of the galaxies in the red sequence bins (1) and (9) (red and orange squares respectively) and of the blue galaxies in bins (4) and (5) (dark and light triangles respectively). The dark grey semi-circles of 1', 2' and 3' radius centered on each clump correspond to regions (18) to (23). The light grey contours show the soft X-ray emission of the hot intracluster gas, as observed with the Chandra space observatory (Maughan et al. [2003]), and were smoothed by applying a Gaussian filter with a FWHM of 2". The field is centered at 01h 52m 41.74s and -13° 57' 52.47".

5.2 Stellar population modeling

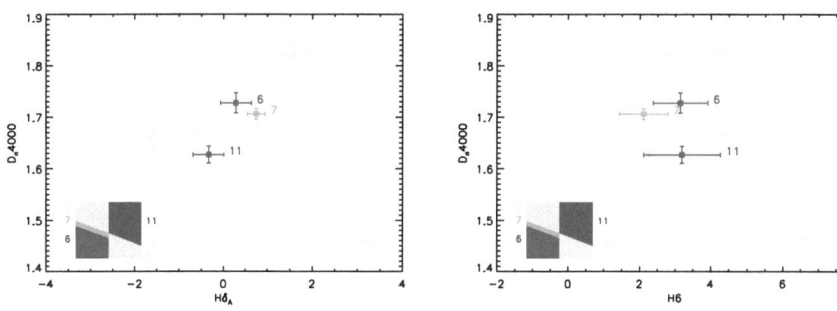

Figure 5.13: $D_n(4000)$ vs Hδ_A (left) and $D_n(4000)$ vs H6 (right) for the composite spectra of the color selected red sequence bins (6) to (11). The insert shows the positions of the red sequence bins relative to each other.

($r_{625} - K_s > 2.3$ and EW([OII])> -5Å) does not distinguish between genuinely passive galaxies and reddened star-forming galaxies with their [OII]λ3727 emission completely suppressed by dust (e.g. Smail et al. [1999]). In Section 5.3, we consider the possibility of dusty starbursts among our passive galaxy sample.

5.2.2 Star formation histories as a function of mass

In Fig. 5.14, we show the star formation weighted age t_{SFR} and final formation look-back time $T - t_{fin}$, from $z = 0.837$, of the best fitting models to the stacked spectrophotometric data of the mass selected bins (12), (13) and (14). As with the luminosity selected bins (1) to (3), we found a clear "downsizing" effect (e.g. Cowie et al. [1996]), the star formation time scales correlating with the average galaxy mass, with the galaxy population of the most massive bin, (12), having formed the bulk of its stars at $z \sim 4$ and the star formation stopping at $z \sim 2$. On the other hand, the star formation histories of the best fitting models to the lowest mass bin, (14), show a delay of ~ 2 Gyr, with a formation redshift z_f of ~ 2 and a final formation redshift z_{fin} of ~ 1. Specifically, we found that $t_{SFR} \propto 2.3 \log M^\star_{phot}$ and $t_{fin} \propto 3.5 \log M^\star_{phot}$, in accord with the results of the fit to the luminosity selected bins (see Fig. 5.15). This is not surprising as the K_s-band luminosity is a good tracer of the stellar mass (see Chapter 3 and Fig. 4.2).

In Fig. 5.15, we plot the mean star formation weighted age t_{SFR} of the best fitting models to the red sequence bins as a function of their mean galaxy mass, compared to the relation obtained from the mass selected bins. As the average mass in the faint blue red sequence bin, (9), is consistent with the other faint red sequence middle and red bins, (10) and (11), we conclude that the age difference found for bin (9) with respect to the other red sequence bins is not due (solely) to an effect of mass but that non-intrinsic factors such as the cluster environment must be taken into account.

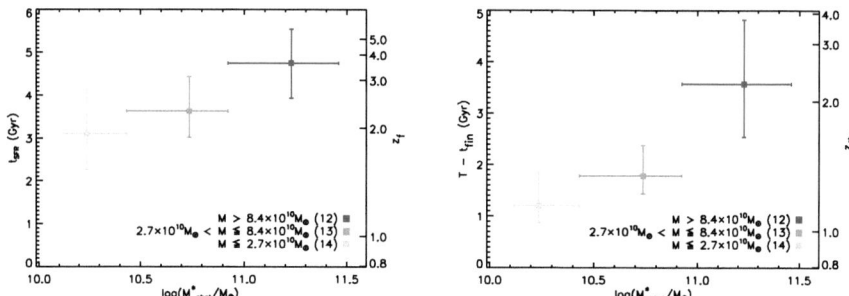

Figure 5.14: Ages of red sequence galaxies as a function of stellar mass: mean and spread of the χ^2-weighted distribution of star formation weighted ages t_{SFR} (left) and final formation look-back times $T - t_{fin}$ (right) of best fitting models to the spectrophotometric data of the mass selected bins (12), (13) and (14).

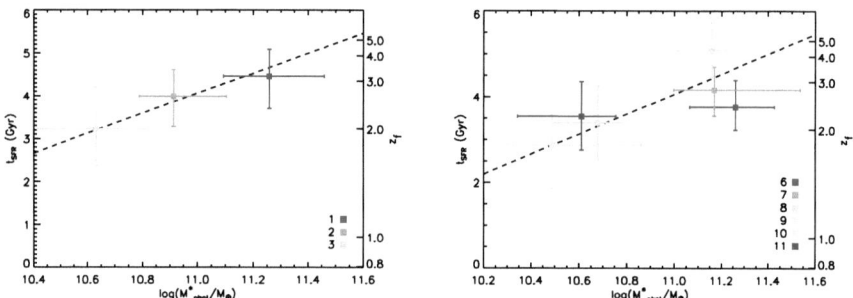

Figure 5.15: Ages of luminosity and color selected red sequence galaxies as a function of stellar mass: mean and spread of the χ^2-weighted distribution of star formation weighted ages t_{SFR} of best fitting models to the spectrophotometric data of the red sequence bins as a function of the mean photometric stellar mass of their galaxies. The dashed line shows the t_{SFR} vs $M^\star_{phot}$ relation derived from the mass selected bins.

5.2 Stellar population modeling

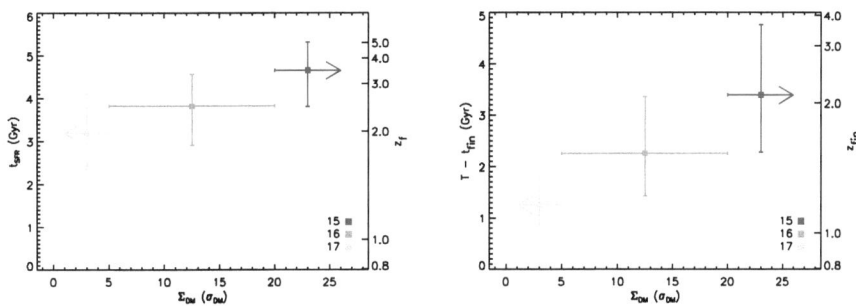

Figure 5.16: Ages of red sequence galaxies as a function of local dark matter density: mean and spread of the χ^2-weighted distribution of star formation weighted ages t_{SFR} (left) and final formation look-back times $T - t_{fin}$ (right) of best fitting models to the to the spectrophotometric data of the dark matter density regions (15), (16) and (17).

5.2.3 Star formation histories as a function of environment

In Fig. 5.16 and 5.17, we show the star formation weighted age t_{SFR} and final formation look-back time $T - t_{fin}$, from $z = 0.837$, of the best fitting models to the stacked spectrophotometric data of the local dark matter density regions (15) to (17) and semi-annuli (18) to (22) around the center of both clumps. As only one passive galaxy was found in the south outermost region (23), with a low S/N spectrum, we ignored it. We found a difference of ~ 1.5 Gyr in age and ~ 2 Gyr in final formation time between the regions with the highest and lowest projected dark matter density respectively. This shows that the star formation timescale was much shorter in the highest dark matter density region, (15), with the galaxy population forming the bulk of its stars at $z > 3$ and the star formation shutting off at $z \sim 2$. On the other hand, the passive galaxy population in the lowest density region, (17), had a more protracted star formation history, with formation and final formation redshifts of ~ 2 and ~ 1 respectively. This is very similar to the star formation parameters derived from the mass selected bins and indeed we see in Fig. 5.19 that a majority of the most massive galaxies are clustered in the central, denser regions of RX J0152.7-1357 while the least massive ones tend to be found in the low dark matter density region (17). Interestingly, the passive galaxies at the center of the northern clump are all from the high mass bin, (12), while the center of the southern clump has a mix of high and intermediate mass galaxies.

Similarly, we found a clear correlation between the star formation weighted age and final formation time and the angular distance to the center of the nearest clump. For the innermost 1' from the northern clump (region (18)), we found a mean star formation weighted age of ~ 4.5 Gyr, corresponding to a formation redshift of ~ 3.5, and final formation look-back time of ~ 3, corresponding to a final formation redshift of ~ 2. Between 1' and 2' from the northern clump's center (region (19)), we found a mean star formation weighted

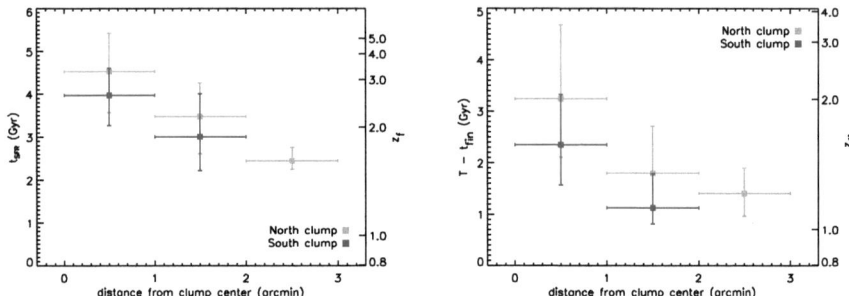

Figure 5.17: Ages of red sequence galaxies as a function of angular distance from the center: mean and spread of the χ^2-weighted distribution of star formation weighted ages t_{SFR} (left) and final formation look-back times $T - t_{fin}$ (right) of best fitting models to the spectrophotometric data of the angular position regions (18) to (20) of the northern clump and (21) and (22) of the southern clump.

age of $\sim$ 3.5 Gyr and at $> 2'$ from the clump center (region (20)) $\sim$ 2.5 Gyr. For the southern clump, we found the same decrease of 1 Gyr per 1' from the center. However, the mean ages of the southern clump are 0.5 Gyr lower compared to the northern one, i.e. 4 and 3 Gyr for regions (21) and (22), corresponding to formation redshifts of 2.6 and 1.8 respectively. The final formation look-back times are 2.4 and 1.1 Gyr, corresponding to redshifts of 1.5 and $\sim$ 1. This is consistent with the passive galaxy population of the southern clump being composed of lower mass galaxies compared to the northern clump (see Fig. 5.19). Finally, we note that this age gradient is comparable to that found in the Lynx supercluster at $z \sim 1.26$ by Mei et al. ([2006b]).

While the cores of both clumps are strongly dominated by high and intermediate mass galaxies, the low density regions of the cluster contain a mix of low to high mass galaxies, as can be seen in Fig. 5.19. This provides a way to distinguish between the effects of mass and environment on galaxy ages. As can be seen in Fig. 5.18, the outer regions of the cluster, where environmental effects are expected to be less important with respect to mass dependent ones, show lower ages with respect to the mass of their galaxies than those derived from the mass-selected bins. This suggests that the ages of the massive, central galaxies are higher than what a pure mass dependent evolution would produce and that the age difference between the central and outer regions of the cluster is consistent with an environment dependent evolution.

5.2 Stellar population modeling

Bin	t_{SFR} (Gyr)	z_f	$T - t_{fin}$ (Gyr)	z_{fin}	Comments
1	$4.5^{+0.6}_{-0.8}$	$3.3^{+1.4}_{-0.8}$	$3.2^{+1.0}_{-0.9}$	$1.9^{+1.1}_{-0.4}$	Bright red sequence
2	$4.0^{+0.6}_{-0.7}$	$2.7^{+0.9}_{-0.6}$	$2.5^{+0.9}_{-0.7}$	$1.5^{+0.6}_{-0.2}$	Middle red sequence
3	$3.2^{+1.0}_{-0.9}$	$2.0^{+1.0}_{-0.4}$	$1.3^{+0.6}_{-0.4}$	$1.1^{+0.2}_{-0.1}$	Faint red sequence
4	$1.9^{+0.3}_{-0.4}$	$1.4^{+0.1}_{-0.1}$	$0.0^{+0.1}_{-0.0}$	$0.84^{+0.1}_{-0.0}$	Bright blue cloud
5	$1.8^{+0.2}_{-0.2}$	$1.3^{+0.1}_{-0.1}$	$0.1^{+0.1}_{-0.1}$	$0.85^{+0.1}_{-0.0}$	Faint blue cloud
6	$3.8^{+0.6}_{-0.5}$	$2.4^{+0.8}_{-0.4}$	$2.3^{+0.8}_{-0.6}$	$1.5^{+0.4}_{-0.2}$	Bright, blue red sequence
7	$4.2^{+0.5}_{-0.6}$	$2.9^{+0.8}_{-0.6}$	$2.7^{+0.9}_{-0.7}$	$1.7^{+0.7}_{-0.3}$	Bright, middle red sequence
8	$5.1^{+0.6}_{-0.6}$	$> 4.4_{-1.1}$	$4.0^{+1.1}_{-1.0}$	$> 2.5_{-0.6}$	Bright, red red sequence
9	$2.9^{+1.1}_{-1.0}$	$1.8^{+0.9}_{-0.4}$	$0.8^{+0.5}_{-0.2}$	$1.0^{+0.1}_{-0.0}$	Faint, blue red sequence
10	$3.4^{+0.9}_{-0.9}$	$2.1^{+1.0}_{-0.4}$	$1.6^{+0.7}_{-0.5}$	$1.2^{+0.3}_{-0.1}$	Faint, middle red sequence
11	$3.5^{+0.8}_{-0.8}$	$2.2^{+0.9}_{-0.4}$	$1.5^{+0.6}_{-0.4}$	$1.2^{+0.2}_{-0.1}$	Faint, red red sequence
12	$4.8^{+0.8}_{-0.8}$	$3.7^{+6.4}_{-1.0}$	$3.6^{+1.3}_{-1.0}$	$2.1^{+3.1}_{-0.5}$	High mass red sequence
13	$3.6^{+0.8}_{-0.6}$	$2.2^{+1.0}_{-0.3}$	$1.8^{+0.6}_{-0.4}$	$1.3^{+0.3}_{-0.1}$	Intermediate mass red sequence
14	$3.1^{+1.0}_{-0.8}$	$1.8^{+1.0}_{-0.3}$	$1.2^{+0.7}_{-0.3}$	$1.1^{+0.2}_{-0.1}$	Low mass red sequence
15	$4.7^{+0.8}_{-0.9}$	$> 3.5_{-1.0}$	$3.4^{+1.4}_{-1.1}$	$> 2.0_{-0.5}$	High local density
16	$3.8^{+0.8}_{-0.9}$	$2.6^{+0.9}_{-0.7}$	$2.3^{+1.1}_{-0.8}$	$1.5^{+0.7}_{-0.3}$	Intermediate local density
17	$3.2^{+1.0}_{-0.8}$	$1.9^{+0.9}_{-0.4}$	$1.3^{+0.6}_{-0.4}$	$1.1^{+0.2}_{-0.1}$	Low local density
18	$4.5^{+0.9}_{-1.0}$	$> 3.4_{-1.0}$	$3.2^{+1.4}_{-1.1}$	$> 1.9_{-0.5}$	North central sector
19	$3.5^{+0.8}_{-0.9}$	$2.2^{+0.8}_{-0.5}$	$1.8^{+0.9}_{-0.7}$	$1.3^{+0.4}_{-0.2}$	North first sector
20	$2.5^{+0.3}_{-0.2}$	$1.6^{+0.2}_{-0.1}$	$1.4^{+0.5}_{-0.4}$	$1.2^{+0.2}_{-0.1}$	North second sector
21	$4.0^{+0.6}_{-0.6}$	$2.6^{+0.9}_{-0.5}$	$2.4^{+1.0}_{-0.8}$	$1.5^{+0.6}_{-0.3}$	South central sector
22	$3.0^{+1.0}_{-0.8}$	$1.8^{+0.9}_{-0.3}$	$1.1^{+0.7}_{-0.3}$	$1.1^{+0.2}_{-0.1}$	South first sector

Table 5.2: Mean star formation weighted age, formation redshift, final formation look-back time from $z = 0.837$ and final formation redshift, with errors, of the models within the 3σ confidence regions of the fits to the average SED and spectrum of each bin.

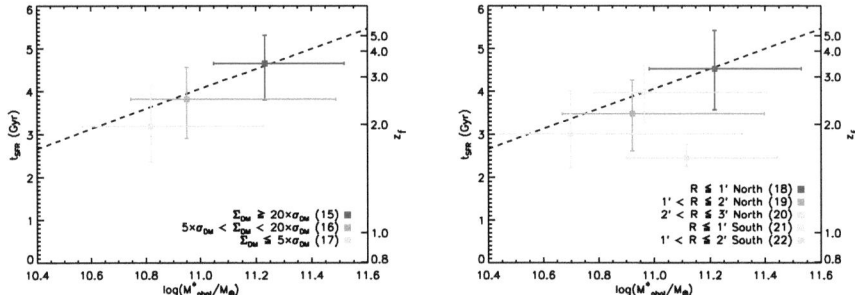

Figure 5.18: Ages of environment selected red sequence galaxies as a function of stellar mass: mean and spread of the χ^2-weighted distribution of star formation weighted ages t_{SFR} of best fitting models to the spectrophotometric data of the dark matter density (left) and angular position (right) regions as a function of the mean photometric stellar mass of their galaxies. The dashed line shows the t_{SFR} vs $M^\star_{phot}$ relation derived from the mass selected bins.

5.3 Effect of metallicity and dust

It is well known that the metal content of galaxies, in particular early-type galaxies, correlates with their mass (e.g. Tremonti et al. [2004], Bernardi et al. [2005], Thomas et al. [2005], Sánchez-Blázquez, et al. [2006]). Indeed, it appears that the color-magnitude relation of early-type galaxies in clusters is caused by this mass-metallicity correlation (Kodama et al. [1998]). This effect might introduce a significant bias in our spectrophotometric analysis as high mass galaxies, being more metal rich, would appear older when fitted with solar metallicity models than their lower mass, lower metallicity counterparts (see Chapter 2). Unfortunately, our composite spectra do not cover the features commonly used to estimate the metallicity of a stellar population, such as Mg_2 or Mgb, and the S/N in their rest-frame optical part is still too low to allow us to constrain the metal content of our different bins using metallicity-sensitive indices. On the other hand, the absence of observed change in the slope of the red sequence from $z \sim 1$ to the present (e.g. Kodama et al. [1998], Blakeslee et al. [2003]) and the fact that the luminous red sequence of the cluster appears already complete suggest that the metallicity of early-type red sequence galaxies didn't change significantly between $z = 0.837$ and $z = 0$. Therefore we can use the low-redshift mass-metallicity relation for early-type galaxies to estimate the metallicity difference between our mass selected bins. From the [Z/H]-$M_\star$ relation of Thomas et al. ([2005]), we found a metallicity of 1.2 and 1.6 solar for the low and high mass bins respectively. This corresponds to an age difference of ~ 0.3 Gyr when using solar metallicity models (see Chapter 2), which is much lower than the ~ 2 Gyr observed difference. We can therefore conclude that metallicity effects do not significantly affect the results of our

5.3 Effect of metallicity and dust

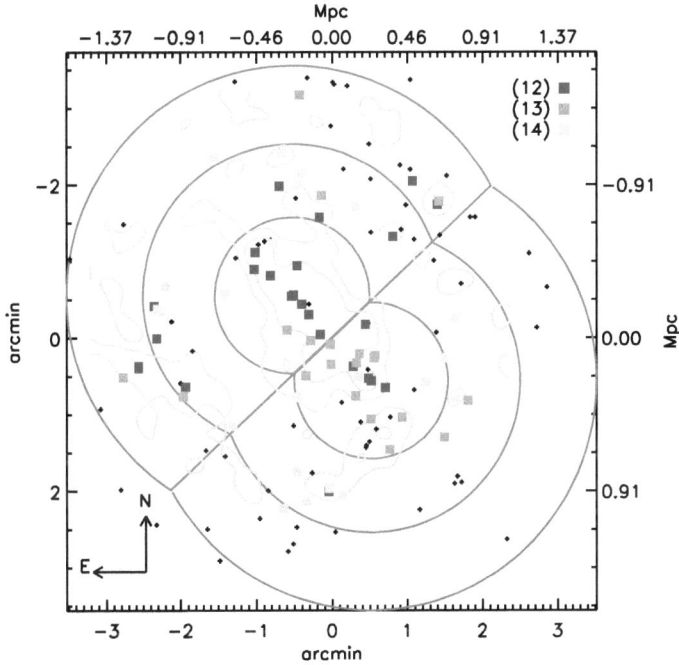

Figure 5.19: Positions on the sky of the galaxies in the mass selected (12), (13) and (14) (blue, red and green respectively). The dark grey semi-circles of 1', 2' and 3' radius centered on each clump correspond to regions (18) to (23). The light grey contours correspond to projected local dark matter densities of 5 and 20 times the critical dark matter density of the cluster. The field is centered at 01h 52m 41.74s and -13° 57' 52.47".

spectrophotometric analysis.

Another issue that must be addressed is the possible presence of dust-enshrouded star forming galaxies in our red sequence bins. As stated in Section 5.2, the [OII]λ3727 emission feature can be greatly affected, if not suppressed, by dust absorption (e.g. Smail et al. [1999]). Therefore, the absence of the [OII]λ3727 feature in the spectra of red sequence galaxies does not guarantee that our red sequence bins be free of dust-enshrouded star forming galaxies. This is especially relevant to the analysis of the faint blue red sequence bin, (9), as it could mean that its composite spectrum is not that of a post-starburst/post-star-forming galaxy population but rather the simple combination of the spectrum of an old passive stellar population and that of a continuously star forming one. We test this using extinction values derived from independent estimates of the star formation rate in RX J0152.7-1357 obtained from infrared observations. As the [OII]λ3727 feature is affected by dust, the EW([OII]) value of our emission line galaxies only provides a lower estimate of the overall star formation rate of late-type galaxies in RX J0152.7-1357. As a truer estimate of the star formation in this cluster, we adopted the value of $22^{+40}_{-10} M_\odot yr^{-1}$ derived by Marcillac et al. ([2007]) from deep 24μm observations with MIPS (Rieke et al. [2004]) on the Spitzer space telescope. We used the Kennicutt ([1998]) relation between the [OII]λ3727 emission and the star formation rate

$$SFR(M_\odot yr^{-1}) = (1.4 \pm 0.4) \times 10^{-41} L([OII])(ergs^{-1}), \tag{5.3}$$

where $L([OII]) \sim (1.4 \pm 0.3) \times 10^{29} \frac{L_B}{L_{B,\odot}} EW([OII]) ergs^{-1}$ is the [OII]λ3727 luminosity as a function of the rest-frame B-band luminosity and [OII]λ3727 equivalent width, and assumed the extinction curve derived by Calzetti et al. ([2000]) for star forming galaxies[2] to estimate the range of star formation rates and $E(B-V)$ values needed to produce an observed EW([OII]) of -5Å given the rest-frame B-band luminosity of the red sequence galaxies. This latter value and its uncertainty were obtained from the best fit τ-models to the SEDs of individual galaxies. At $z \sim 0.84$, the rest-frame B band falls between the i_{775} and z_{850} filters. The rest-frame B-band luminosity is therefore strongly constrained by the observed SED and only weakly model-dependent. In Fig. 5.20, we show the relation between the star formation rate and $E(B-V)$, assuming an observed EW([OII]) of -5Å, with the shaded area representing the range of star formation rates estimated from the 24μm observations. We found that the amount of dust needed to damp the [OII]λ3727 down to EW([OII]) = -5Å results in an extinction of at least $E(B-V) = 0.6$.

---

[2] $F(\lambda) = F_0(\lambda) 10^{-0.4 E_s(B-V) k'(\lambda)}$, where $F(\lambda)$ and $F_0(\lambda)$ are the observed and intrinsic continuum fluxes respectively,

$$k'(\lambda) = \begin{cases} 2.659(-1.857 + 1.04/\lambda) + R'_V & 0.63\mu m \leq \lambda \geq 2.20\mu m \\ 2.659(-2.156 + 1.509/\lambda - 0.198/\lambda^2 + 0.011/\lambda^3) + R'_V & 0.12\mu m \leq \lambda < 0.63\mu m \end{cases}$$

and $R'_V = 4.05 \pm 0.80$

5.3 Effect of metallicity and dust

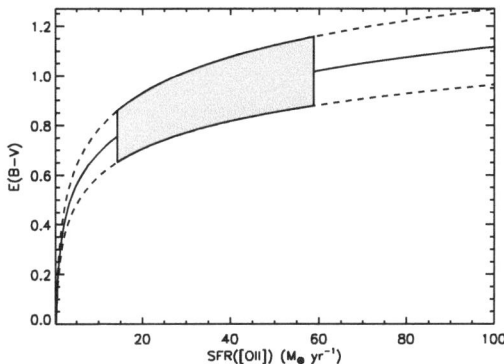

Figure 5.20: $E(B-V)$ as a function of SFR([OII]), assuming an observed EW([OII]) of -5 Å (solid line). The dashed lines represent the error due to the uncertainty on the effective obscuration R'_V (Calzetti et al. [2000]) and the rest-frame B-band luminosity scatter of the red sequence galaxies of RX J0152.7-1357. The shaded area show the range of values derived from the Marcillac et al. ([2007]) estimate of the star formation rate in the cluster. A significant amount of dust ($E(B-V) \gtrsim 0.6$) is needed to suppress the [OII]λ3727 emission.

In a second time, we extended the grid of Bruzual & Charlot ([2003]) solar metallicity models down to ages of 0.1 Myr and reddened these model spectra using the Calzetti et al. ([2000]) prescription, assuming the putative extinction derived from the limiting EW([OII]) of -5Å (see above) and the Marcillac et al. ([2007]) SFR value. We found that only models younger than 7 Myr could reproduce the observed red sequence colors when reddened. We then compared this subset of models to the spectra of galaxies in the bright red and faint blue red sequence bins, (8) and (9), using the $D_n(4000)$ and Hδ_A indices as well as the equivalent width of H6. We found that the reddened models could reproduce the equivalent width of the Balmer features of some of the galaxies but that the 4000 Å break is consistently shallower (see Fig. 5.21). The spectral features and color of one, possibly two, galaxies in the bright red sequence bin (8) are consistent with a dust-enshrouded star forming galaxy but there are none in the faint blue red sequence bin (9). We conclude that dusty star forming galaxies do not form a significant fraction of the population of our red sequence bins but rather that the galaxy population of the faint blue end of the red sequence is truly a post-starburst/post-star-forming one.

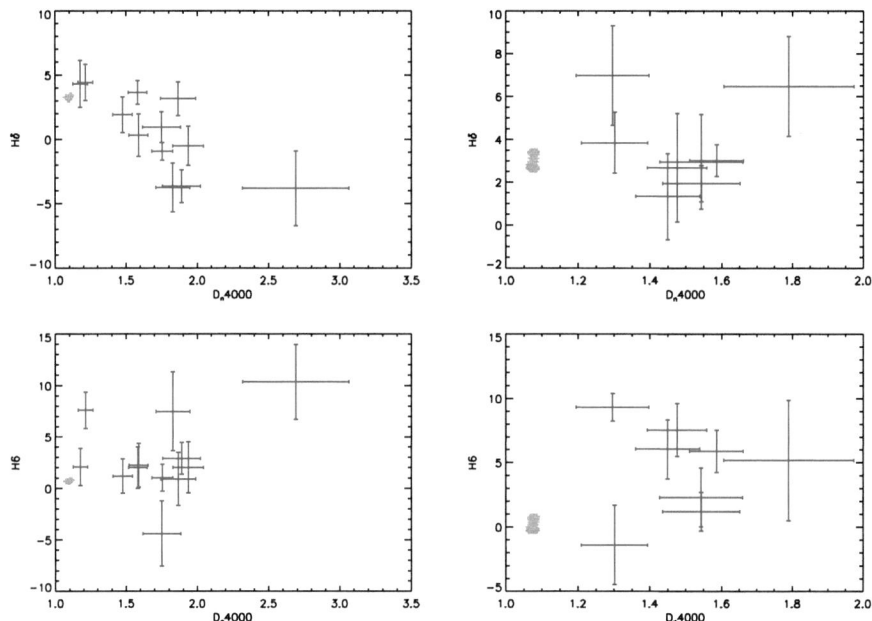

Figure 5.21: Spectral indices of red sequence galaxies compared to dust-reddened models that reproduce the galaxies' colors: distribution of Hδ_A (top) and H6 (bottom), as a function of D$_n$4000, of the galaxies in the bright red and faint blue red sequence bins, (8) (left) and (9) (right), in blue. The index values of the models which reproduce the observed $r - K_s$, assuming extinction values consistent with the observed L([OII]) and 24μm-derived SFR, are shown in red.

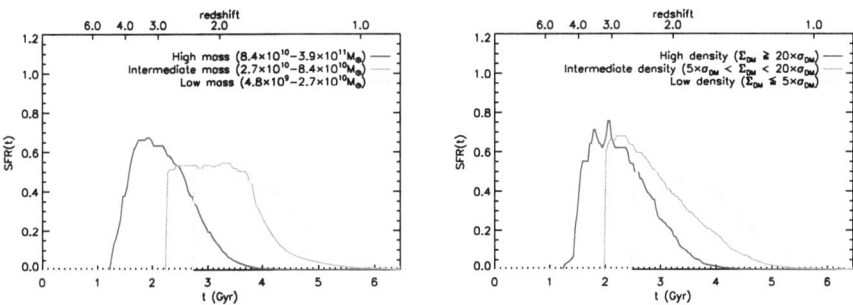

Figure 5.22: Median star formation history of the best fitting models to the composite spectrophotometric data of the mass-selected bins and dark matter density regions. The lower x-axis shows the cosmic time t and the upper x-axis the corresponding redshift.

5.4 Summary

In this Chapter, we have used a set of 134 galaxy spectra and 5-band photometry to investigate the stellar population properties of early-type galaxies in the $z = 0.84$ cluster RX J0152.7-1357. This very rich dataset allowed us to study the variation of star formation histories within the cluster as a function of galaxy luminosity, color and photometric stellar mass, as well as environmental density. By modeling the spectrophotometric properties of early-type galaxies, we found a clear correlation between age, mass (or luminosity) and position with respect to the cluster center. RX J0152.7-1357 consists of two components in the process of merging and each core is populated by massive, old early-type galaxies. We found that these have formed the bulk of their stars at $z > 3$ and stopped star formation at $z \sim 2$ altogether, with the most massive galaxies ($> 8 \times 10^{10} M_\odot$) being ~ 1 Gyr older than less massive ones ($> 3 \times 10^{10} M_\odot$). These ages suggest a formation scenario involving an accelerated star formation history and early quenching of star formation, followed possibly by further mass assembly via dry mergers, and appear to be consistent with studies of higher redshift clusters (Blakeslee et al. [2003], Holden et al. [2004], Lidman et al. [2004], Mei et al. [2006a], Tanaka et al. [2008]). This is illustrated in Fig. 5.22, where we plot the median star formation history of best fitting models as a function of galaxy mass and local dark matter density. At the faint end of our sample, however, we found a population of galaxies that, while now passive, shows sign of having only recently ($\lesssim 1$ Gyr from the cluster's epoch) stopped star formation. We also used our stellar population modeling to test for and reject the presence of dusty star forming galaxies among our early-type galaxy sample.

While stellar mass roughly follows environmental density (see Fig. 5.19), with the most massive galaxies at the cluster's center and the least massive ones in the periphery, we found a difference between the age-mass and age-environment relations which permitted us to partially distinguish between the effects on the evolution of cluster early-types due to the "intrinsic" property of mass and those due to the cluster environment. Interestingly, we found that the southern component of the cluster is overall ∼ 0.5 Gyr younger than the northern one, consistently with the southern clump being a less massive system (Demarco et al. [2005], Blakeslee et al. [2006]). Overall, the picture emerging from the spectrophotometric analysis is one where low mass galaxies on the cluster rim are transformed from blue star forming to red passive galaxies as they fall into the denser core. The lack of correlation between the distribution of newly passive galaxies and the X-ray emission implies that the quenching of star formation did not occur through interaction with the dense hot intracluster medium but rather by galaxy-galaxy interactions. The downsizing of star formation in cluster galaxies appears thus to be an environmental effect, with the star formation time scale being shorter than the assembly one (i.e. galaxies migrate from the blue cloud to the the faint end of the red sequence, with most of the mass assembly happening later along the red sequence). On the other hand, the high ages of the brightest early-type galaxies are not incompatible with a "monolithic collapse" scenario and it is likely that the quenching of star formation happens through different mechanisms according to the epoch and the environment (e.g. Cooper et al. [2007], Faber et al. [2007]).

Chapter 6

Moving to higher redshifts: two clusters at $z \sim 1$ and $z \sim 1.4$

In the previous Chapters, we have studied a relaxed cluster at $z = 1.24$, RDCS J1252.9-2927, and a more complex one at $z = 0.84$, RX J0152.7-1357. While we found that the oldest galaxies in both clusters have formed at $z > 3$, the star formation histories of galaxies in both clusters are hardly comparable. First because RX J0152.7-1357 is in a merging phase, which would be expected to result in an influx of young early-type galaxies to the cluster population, and secondly because we expect a different "progenitor bias" (van Dokkum & Franx [2001b]), as the two clusters are observed at significantly different redshifts. It is therefore difficult to draw conclusions on the general history of star formation in galaxy clusters from such a small and heterogeneous sample. In order to better understand the processes that drive galaxy evolution in dense environments, a statistical study of a larger sample of high redshift galaxy clusters is needed. This may soon become feasible, as ongoing or upcoming surveys, such as those making use of the Sunyaev Zel'dovich effect (e.g. Carlstrom, Holder & Reese [2002]), promise to increase the number of confirmed high-redshift clusters by at least an order of magnitude. In this Chapter, we apply the spectrophotometric method described in Chapter 2 to two distant X-ray selected clusters, XMMU J1229+0151 at $z = 0.98$ and XMMU J2235.3-2557 at $z = 1.39$, both discovered in the XMM-Newton Distant Cluster Project (XDCP; Böhringer et al. [2005], Fassbender et al. [2008]), and compare the results of this analysis with those of the previous Chapters.

This Chapter is organized as follows. In Section 6.1, we describe the sample of early-type galaxies in XMMU J1229+0151 and the results of the spectrophotometric fit. In Section 6.2, we describe the sample of early-type galaxies of XMMU J2235.3-2557 and the fit to its spectrophotometric data.

6.1 The cluster XMMU J1229+0151 at $z = 0.98$

The first cluster is XMMU J1229+0151, an X-ray selected cluster at $z = 0.975$ detected by the XDCP. Results of the ACS and XMM observations of this cluster are reported in Santos et al. ([2009]). Here we describe the modeling of the spectrophotometric data for which we were responsible.

6.1.1 Observations and sample selection

Observations of the cluster were carried out in the i_{775} and z_{850} bands using the *HST*/ACS Wide Field Camera and in the J and K_s bands using SofI on the ESO NTT. The data were acquired in the framework of the Supernova Cosmology Project in December 2006 with total exposure times of 4110 and 10940 seconds respectively. The SofI images were taken in March 2007 with total exposure times of 2700 and 3600 seconds for the J and K_s bands respectively. Magnitudes were measured with SExtractor (Bertin & Arnouts [1996]) in apertures of 0.5" for i_{775} and z_{850} and 1.16" for J and K_s, then corrected to 3", which encloses most of the galaxy light. The magnitudes were corrected for galactic extinction using the NED galactic extinction calculator[1]. The extinction corrections are 0.046 mag in i_{775}, 0.036 in z_{850}, 0.021 in J and 0.009 in K_s. A color-composite image of the central region of XMMU J1229+0151 is shown in Fig. 6.1.

In addition, spectra of 74 objects in the field of the cluster were taken with FORS2 on the ESO VLT, also in the framework of the Supernova Cosmology Project, using the 300I grism (see Chapter 4), with 4 pointings in MOS mode (movable slits) and one in MXU mode (precut mask). This yielded 64 redshifts, of which 27 belong to the cluster. The latter were visually classified by J. Santos using the scheme of Postman et al. ([2005]). The morphology of the cluster members is clearly dominated by elliptical galaxies, with only one member classified as a spiral galaxy. Of the 27 spectroscopic members, 20 are common to the SofI and ACS images. In addition, four of these 20 spectroscopic members are found in two red galaxy pairs that are not properly resolved in the SofI images. Fig. 6.2 shows the spatial distribution of the different cluster members. Four of the remaining 16 galaxies show [OII]λ3727 emission, a telltale sign of star formation, and two of them also have an [OIII]λ4959,5007 emission feature. One of them is the (edge-on) spiral while the other three had been classified as elliptical galaxies. The remaining 12 passive galaxies all lie on the red sequence of the cluster, as shown in Fig. 6.3. Fig. 6.4 shows the spectra of the 12 confirmed passive cluster members. Of those twelve passive galaxies, one has a moderately deep Hδ absorption feature (EW(Hδ)> 3Å) and three show strong (EW(Hδ)> 8Å) Balmer absorption, indicating a recent episode of star formation. We refer to them as "k+a" and "a+k" types respectively, following Dressler et al. ([1999]). As in RX J0152.7-1357, the k+a/a+k and emission line galaxies tend to avoid the cluster center (see Fig. 6.2). We selected as our early-type galaxy sample (hereafter XMM 1229) the 8 remaining passive

---
[1] http://nedwww.ipac.caltech.edu/

6.1 The cluster XMMU J1229+0151 at $z = 0.98$

Figure 6.1: 1.5'×1.5' ($\sim$ 720 kpc at $z = 0.98$) region centered on XMMU J1229+0151, from a color composite image of HST/ACS i_{775} and z_{850} (Santos et al. [2009]).

members which showed little Balmer absorption, hereafter referred to as "k" type. In Table 6.1, we show the relevant properties of the 12 passive spectroscopic members. Additionally, we considered a sample of galaxies without spectroscopic redshifts but classified as E or S0 and lying on the red sequence defined by the spectroscopic members (see Fig. 6.3). We found 42 such "early-type" galaxies in the color-magnitude diagram of the ACS bands, of which only 18 are resolved in the SofI data.

6.1.2 Modeling the star formation history

Initially, we computed photometric stellar masses for the 12 passive spectroscopic members and the 18 morphologically selected galaxies on the red sequence, by fitting their 4-band SEDs with a grid of τ-models. From this modeling, we also derive ages for the individual galaxies. As in the previous Chapters, this grid was built from Bruzual & Charlot ([2003]) solar metallicity templates, assuming a Salpeter ([1955]) IMF and a delayed, exponentially declining star formation rate. The values of T ranged from 200 Myr to 5.8 Gyr, the age of the Universe at $z = 0.975$, with increments of 250 Myr, and τ from 0 to 2 Gyr with increments of 50 Myr. In Fig. 6.5, we show the response functions of the two ACS and two SofI filters overlayed on the best fit model to one of the three brightest cluster galaxies. In Fig. 6.6, we show the distribution of photometric stellar masses of the 12 passive spectroscopic members (in color) and of the 18 morphologically selected galaxies on the red sequence (in black).

As in Chapter 4 and 5, we also coadded the spectra of the 8 "k" galaxies (see Fig. 6.2) as well as their photometry and proceeded to compare the composite SED and spectrum with the grid of composite stellar population models. For the fit to the composite spectrum (S/N$\sim$16), we ignored a 100 Å interval centered on the atmospheric A-band line at $\sim$ 7600 Å, which had not been properly subtracted in the individual spectra (see Fig. 6.8), and considered a 500 Å interval centered at 4000 Å. Fig. 6.7 shows the confidence regions of the fits to the composite SED and spectrum of the XMM 1229 sample. Table 6.2 shows the results of the fit to the stacked spectrophotometric data of the XMM 1229 sample.

Unlike with RDCS J1252.9-2927 and RX J0152.7-1357, where the confidence regions of the fit to the SED and that to the spectrum mostly overlapped, the fits to the stacked SED and spectrum of the XMM 1229 sample give noticeably different results. As shown in Fig. 6.9 and Table 6.2, the mean star-formation weighted age of the best fitting models to the composite SED of the XMM 1229 sample is $\sim$4 Gyr, corresponding to a formation redshift of > 4 and consistent with the values for the massive early-type galaxies in RDCS J1252.9-2927 and RX J0152.7-1357 (see Chapters 4 and 5). On the other hand, the fitting of the composite spectrum only yields younger ages, $t_{SFR} \sim$ 2.7 Gyr with a formation redshift of $\sim$ 1.3. The discrepancy between the ages inferred independently from the SED and spectra could have the following causes: first, it may be that the four bands are not sufficient to accurately determine star formation histories as information on the rest-frame

6.1 The cluster XMMU J1229+0151 at $z = 0.98$

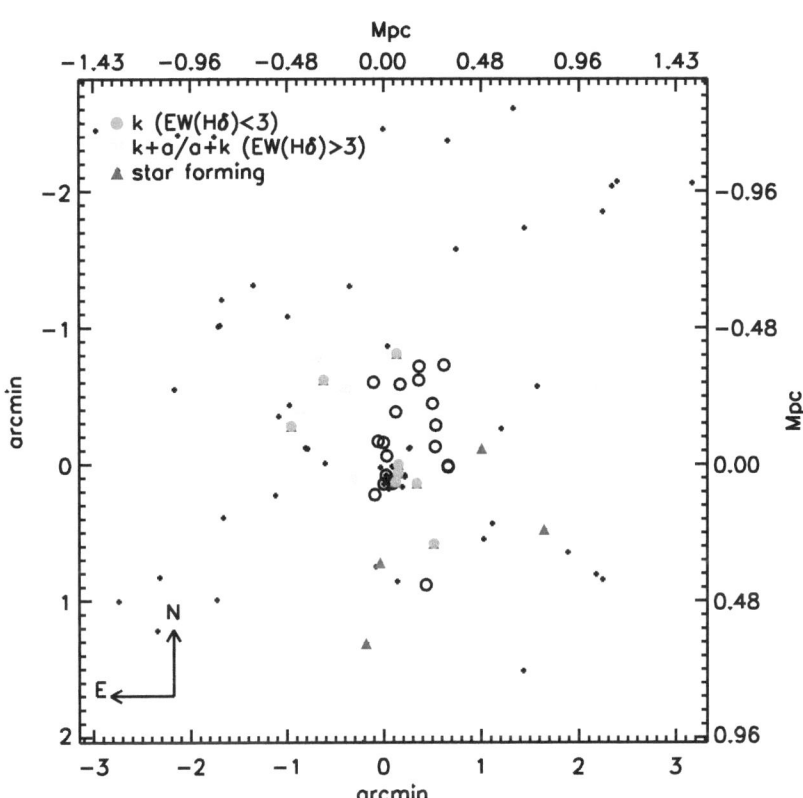

Figure 6.2: Positions of the cluster members (filled symbols) in the field of XMMU J1229+0151. Star forming galaxies, including a spiral galaxy with no apparent [OII] emission, are shown by blue triangles, passive galaxies with EW(Hδ_A)< 3Å by red circles and passive galaxies with EW(Hδ_A)> 3Å (corresponding respectively to the types k and k+a of Dressler et al. [1999]) by orange squares. The open symbols represent the morphologically selected galaxies on the red sequence without spectroscopic redshifts. The field is centered at 12h 29m 29.79s and 01° 51' 71".

ID	RA	DEC	z	i_{775}	z_{850}	J	K_s	class
3428	12h 29m 31.04s	01° 51' 22.81	0.984	23.534±0.017	21.893±0.053	20.895±0.069		a+k
3430	12h 29m 29.29s	01° 51' 21.87	0.974	22.752±0.009	21.804±0.004	20.618±0.017	19.687±0.023	k
3495	12h 29m 29.18s	01° 51' 29.62	0.980	23.551±0.018	21.692±0.009	20.856±0.066		k
3507	12h 29m 29.20s	01° 51' 25.76	0.976	22.636±0.008	21.660±0.004	20.595±0.016	19.647±0.022	k
3524	12h 29m 31.37s	01° 52' 3.63	0.969	23.034±0.014	22.164±0.007	20.853±0.021	19.879±0.027	a+k
4126	12h 29m 33.61s	01° 51' 46.41	0.973	23.084±0.012	22.185±0.006	21.434±0.035	20.729±0.059	k
4155	12h 29m 33.26s	01° 51' 52.22	0.969	23.462±0.016	22.598±0.009	21.899±0.054	20.883±0.068	k+a
5411	12h 29m 29.26s	01° 52' 18.44	0.974	23.757±0.021	22.853±0.011	22.015±0.060	20.915±0.070	k
5417	12h 29m 32.59s	01° 52' 16.57	0.977	23.247±0.013	22.329±0.007	21.419±0.034	20.461±0.046	a+k
5499	12h 29m 32.27s	01° 52' 6.92	0.973	23.258±0.014	22.369±0.007	21.719±0.045	20.552±0.050	k
20013	12h 29m 28.42s	01° 51' 21.33	0.979	23.719±0.021	22.913±0.012	22.047±0.061	21.204±0.091	k
20014	12h 29m 27.71s	01° 50' 54.86	0.969	23.164±0.013	22.355±0.007	21.863±0.052	21.151±0.087	k

Table 6.1: Relevant properties of the 12 passive spectroscopic members of XMMU J1229+0151: right ascension, declination, redshift, magnitudes in the i_{775}, z_{850}, J and K_s bands and spectral class.

6.1 The cluster XMMU J1229+0151 at $z = 0.98$

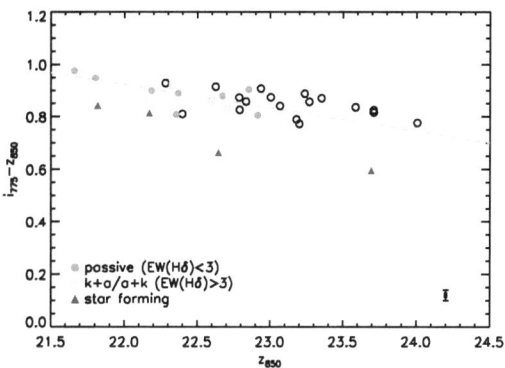

Figure 6.3: Color-magnitude relation of the red galaxies in XMMU J1222+0151, with the confirmed members represented by filled symbols. Star forming galaxies, including a spiral galaxy with no apparent [OII] emission, are shown in blue, passive galaxies with EW(Hδ_A)< 3Å in red and passive galaxies with EW(Hδ_A)> 3Å (corresponding respectively to the types k and k+a of Dressler et al. [1999]) in orange. The open symbols represent the morphologically selected galaxies on the red sequence without a spectroscopic redshift.

UV flux, which could be used to constrain recent star formation, is missing. We also note that the i_{775} and z_{850} filters are not optimal at this redshift, as the i_{775} straddles the 4000 Å break. On the other hand, the discrepancy could be caused by an overestimate of the flux in the SofI bands with respect to the ACS ones. This would skew the fit toward models with higher near-IR fluxes and deeper 4000 Å than what is observed. To examine the first possibility, we performed a new spectrophotometric fit to the composite spectrum and i_{775}, z_{850}, J and K bands of the GOODS and RDCS 1252 samples (see Chapter 4), whose SEDs were well sampled in 8-9 bands. We found that, for the GOODS and RDCS 1252 samples, this does not produce higher ages. Using only the 4-band SEDs, we obtained star formation weighted ages for the cluster and the field that were lower by 0.15 and 0.2 Gyr respectively, and final formation times 0.1 Gyr higher for both samples, and therefore did not find any significant bias. The distributions of these parameters for the best fitting models were also very similar. We then conclude that the second possibility (overestimate of the near-IR fluxes) is more likely. This is supported by the observation that, for five of the eight galaxies in the XMM 1229 sample, the flux in the J and K_s bands of the best fit model to the SED is lower than the observed flux (see Fig. 6.8), thus leading to an overestimate of the ages by ~ 1 Gyr. As an independent constraint, the measured small scatter of the $i_{775} - z_{850}$ red sequence ($\sim$0.04; Santos et al. [2009]) is similar to that in RDCS J1252.9-2927 and therefore consistent with ages of $\sim 3 - 4$ Gyr, as discussed in Chapter 4.

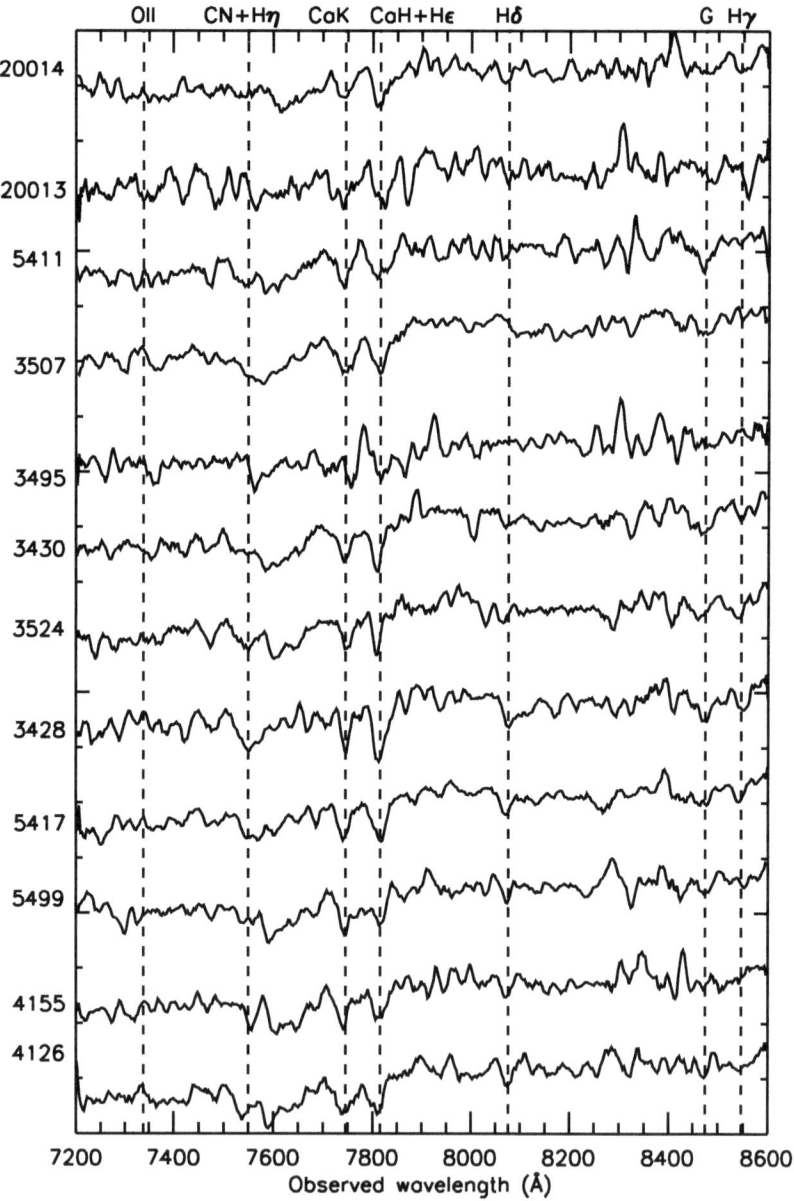

Figure 6.4: Spectra of the 12 passive member galaxies of XMMU J1229+0151, smoothed with a width of 3Å.

6.1 The cluster XMMU J1229+0151 at $z = 0.98$

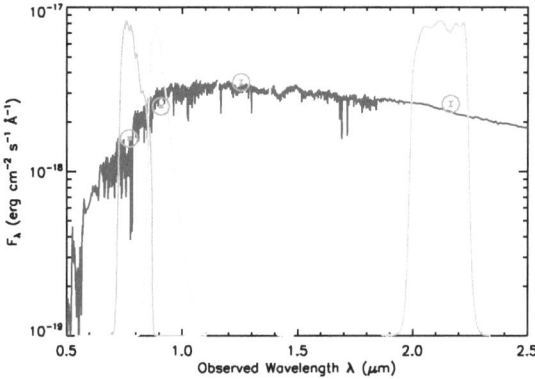

Figure 6.5: SED fit of one of the three brightest cluster galaxies. The flux measurements in the i_{775}, z_{850}, J and K_s bands are represented by red circles and the best fit model to the SED is shown in blue. The filter transmission curves are shown in light blue, green, yellow and red respectively.

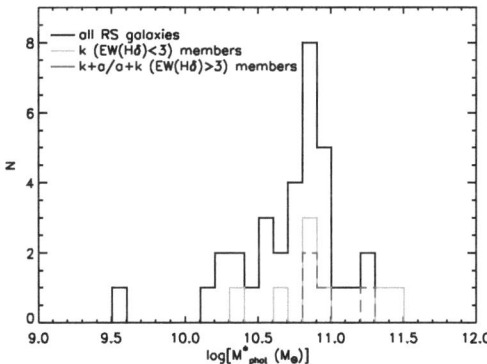

Figure 6.6: Distribution of photometric stellar masses, derived from SED fitting, of the passive member galaxies of XMMU J1229+0151. Passive galaxies with EW(Hδ_A)< 3Å are shown in red and those with EW(Hδ_A)> 3Å in blue (corresponding respectively to the types k and k+a/a+k of Dressler et al. [1999]). The black histogram shows the total distribution of the masses of the 12 spectroscopic members and 18 morphologically selected red sequence galaxies.

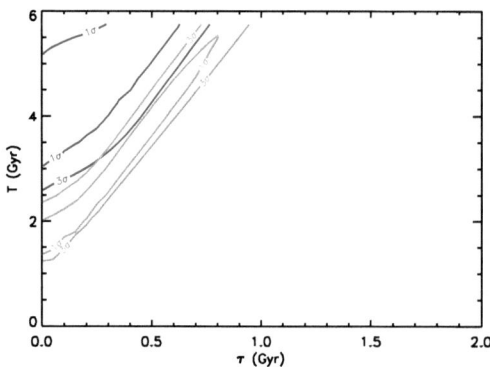

Figure 6.7: 1σ and 3σ confidence regions of the fit to the average SED (blue) and spectrum (red) of the passive "k" type (EW(Hδ)< 3Å) spectroscopic members of XMMU J1229+0151.

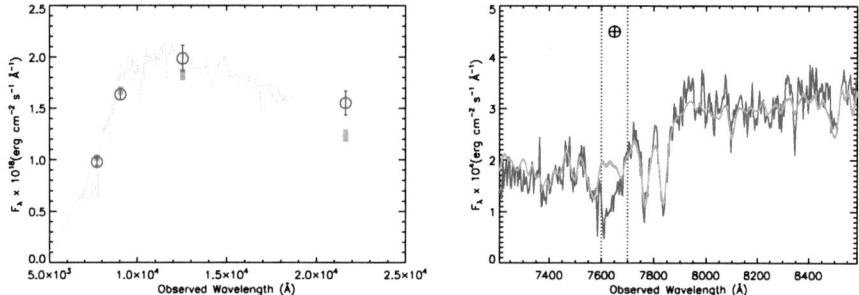

Figure 6.8: Average SED (left) and spectrum (right) of the passive galaxies in the XMM 1229 sample, in blue, and best fit models (red) within the 3σ confidence regions of the fit to the stacked spectrophotometric data. The best fit model to the average SED is shown in green (left) and the telluric line at ∼7600 Å is shown by the dotted lines (right).

Fit to the	t_{SFR} (Gyr)	z_f	$T - t_{fin}$ (Gyr)	z_{fin}
SED+spectrum	$3.7^{+0.4}_{-0.5}$	$3.1^{+0.5}_{-0.5}$	$1.1^{+0.1}_{-0.1}$	$1.3^{+0.03}_{-0.02}$
SED	$4.1^{+0.7}_{-0.8}$	$4.4^{+5.2}_{-1.5}$	$2.9^{+1.3}_{-1.0}$	$2.7^{+3.5}_{-1.0}$
spectrum	$2.7^{+0.8}_{-0.8}$	$2.2^{+0.8}_{-0.5}$	$1.0^{+0.5}_{-0.3}$	$1.3^{+0.2}_{-0.1}$

Table 6.2: Mean star formation weighted age, formation redshift, final formation look-back time from $z = 0.975$ and final formation redshift, with errors, of the best fitting models to the average SED and spectrum of the XMM 1229 sample.

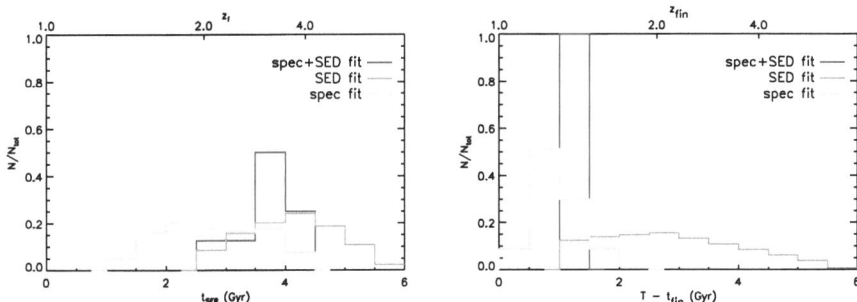

Figure 6.9: Distribution, as a function of the star formation weighted age (left) and lookback time to the final formation time (right), of the best fitting models within the 3σ confidence regions of the fit to the spectrophotometric (blue), photometric (red) and spectroscopic (green) data of the XMM 1229 sample.

As in previous Chapters, we also found clear evidence of "downsizing", with the most massive galaxies being also the oldest. In Fig. 6.10, we show the correlation between the star formation weighted ages and photometric stellar masses for both samples. We note that, as the χ^2 of the fit to the SED of the least massive (but apparently old) morphologically selected galaxy is high, it is likely an interloper. We also compared the best fit t_{SFR}-$M^\star_{phot}$ correlation to the passive galaxies of XMMU J1229+0151 with those of the cluster samples in this Chapter and Chapters 4 and 5. These are plotted in Fig. 6.10 as dashed lines, normalized to that of XMMU J1229+0151 at $\log(M^\star_{phot})$=10.75. Interestingly, there is no significant variation in slope, which implies that the timescales of galaxy evolution are similar in all four clusters. This suggests in turn that either the shutdown of star formation occurs independently of environment but is rather only a function of the mass, which is not very likely in light of the previous Chapters, or that the environments in which the galaxies of a given mass ceased star formation were not very different. This suggests, for example, that the quenching of star formation occurred at roughly the same environmental density in all four clusters (while not necessarily at the same distance from each cluster center).

6.2 The cluster XMMU J2235.3-2557 at $z = 1.39$

The second cluster examined in this Chapter, and last in this work, is XMMU J2235.3-2257 (Mullis et al. [2005]) at $=1.39$. Discovered in the framework of the XMM-Newton Distant Cluster Project (Böhringer et al. [2005]), it is the second most distant confirmed X-ray selected cluster to date and the most massive discovered so far at $z > 1$. Results of the observations of this cluster are presented in Rosati et al. ([2009]). Here we describe the modeling of the spectrophotometric data.

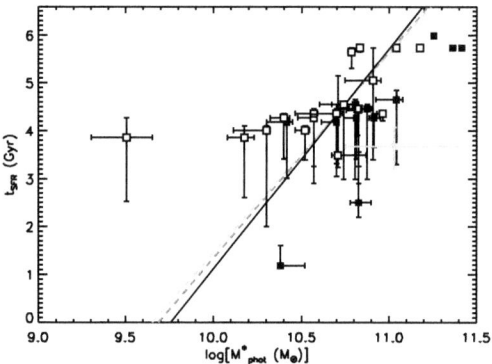

Figure 6.10: Star formation weighted age as a function of the photometric stellar mass, both derived from SED fitting, of the passive spectroscopic cluster members (filled black symbols) and morphologically selected red sequence galaxies (open symbols). The mean mass of the XMM 1229 sample and star formation weighted age from its composite spectrophotometry are shown by the yellow symbol. The best fit correlation is represented by the continuous black line while those of the clusters RX J0152.7-1357, RDCS J1252.9-2927 (see Chapters 5 and 4 respectively) and XMMU J2235.3-2557 are shown by the dashed blue, red and green lines respectively.

6.2 The cluster XMMU J2235.3-2557 at $z = 1.39$

RA	DEC	z	i_{775}	z_{850}	J	K
22h 35m 19.08s	-25° 58' 27.43s	1.390	23.97	23.09	21.83	20.88
22h 35m 20.82s	-25° 57' 40.13s	1.394	23.63	22.76	21.34	20.20
22h 35m 20.92s	-25° 57' 35.89s	1.400	24.39	23.62	22.40	21.03
22h 35m 25.45s	-25° 56' 52.88s	1.389	23.71	22.74	21.37	20.21
22h 35m 22.79s	-25° 56' 25.04s	1.389	24.08	23.22	22.03	21.08
22h 35m 21.59s	-25° 57' 25.60s	1.380	23.69	22.95	22.02	21.18
22h 35m 20.69s	-25° 57' 44.67s	1.375	23.91	23.03	21.73	20.40
22h 35m 16.11s	-25° 57' 1.35s	1.385	24.56	23.50	22.34	21.08
22h 35m 22.74s	-25° 57' 6.06s	1.400	23.58	22.77	21.90	21.09
22h 35m 20.69s	-25° 57' 37.85s	1.383	23.81	22.85	21.52	20.39
22h 35m 26.54s	-25° 57' 50.65s	1.397	24.12	23.16	22.27	21.31
22h 35m 19.03s	-25° 57' 51.59s	1.395	24.48	23.55	22.39	21.15
22h 35m 27.06s	-25° 56' 34.83s	1.386	24.08	23.07	22.12	21.31
22h 35m 16.11s	-25° 57' 1.55s	1.385	24.56	23.50	22.34	21.08
22h 35m 27.06s	-25° 56' 34.83s	1.387	24.08	23.07	22.12	21.31

Table 6.3: Relevant properties of the 15 passive spectroscopic members of XMMU J2235.3-2557: right ascension, declination, redshift and magnitudes in the i_{775}, z_{850}, J and K bands.

6.2.1 Observations and sample selection

XMMU J2235.3-2557 was observed in the i_{775} and z_{850} bands with *HST*/ACS and in the J and K bands with ISAAC on the VLT. The ISAAC observations are a mosaic of 3×3 pointings covering 7.5'×7.5', with exposure times ranging from one hour per filter in the core to 30 minutes in the other frames. For the ACS images, the total exposure times are 5060 sec in i_{775} and 6240 sec in z_{850}. Magnitudes were measured using SExtractor in apertures of 0.75" for the i_{775} and z_{850} and and 1" for J and K, then corrected to 2" and 4" respectively. In both cases, this safely encloses most of the galaxy light. The magnitudes were corrected for galactic extinction using the NED extinction calculator. The extinction corrections are 0.043 mag in i_{775}, 0.032 in z_{850}, 0.019 in J and 0.008 in K. A color-composite image of XMMU J2235.3-2557 is shown in Fig. 6.11.

As part of three observing runs, spectra of 155 objects in the field of the cluster were obtained with *VLT*/FORS2 in MXU mode using the 300I grism. This yielded 29 cluster members, of which 17 are passive (i.e. with EW([OII])> -5Å) and 15 are also found in ACS and ISAAC images. We refer to these 15 passive galaxies as the XMM 2235 sample. In Table 6.3, we show the relevant properties of the 15 passive spectroscopic members. Fig. 6.12 shows the spatial distribution of the different cluster members.

6.2.2 Modeling the star formation history

As in Section 6.1, we computed photometric stellar masses for the 15 passive spectroscopic members in the XMM 2235 sample by fitting their 4-band SEDs with a grid of τ-models.

Figure 6.11: 3'×3' (~ 1.5 Mpc at $z = 1.39$) region centered on XMMU J2235.3-2557, from a color composite image of i_{775}, z_{850} and K (Rosati et al. [2009]). The green contours show the X-ray emission from a deep Chandra exposure.

6.2 The cluster XMMU J2235.3-2557 at $z = 1.39$

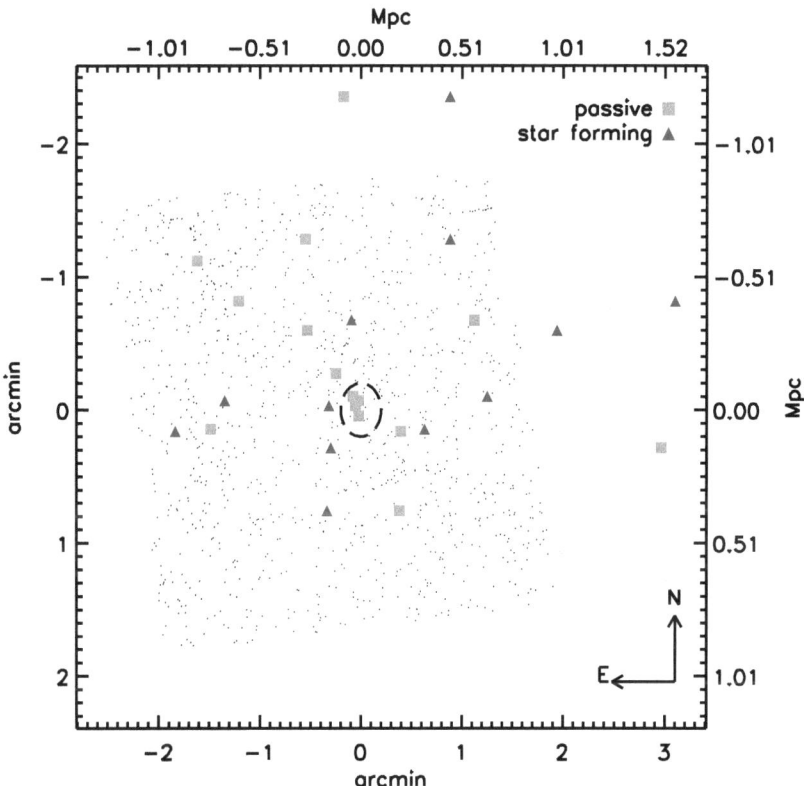

Figure 6.12: Positions of the passive (red squares) and star forming (blue triangles) cluster members in the field of XMMU J2235.3-2257. The black points show the positions of the objects in the photometric catalog. The field is centered at 22h 35m 20.6s and -25° 57' 42". The dashed circle of 12" (∼ 100 kpc) radius shows the four core galaxies in our sample.

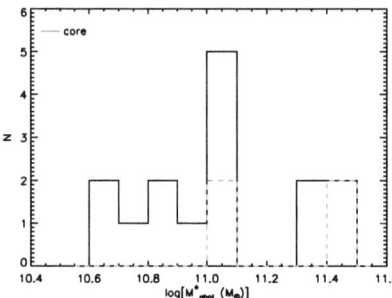 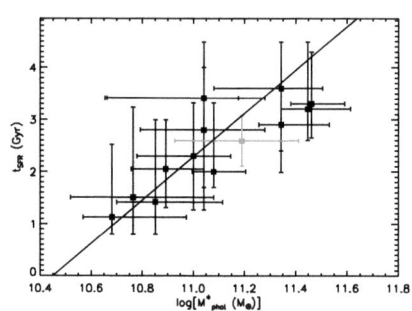

Figure 6.13: Left: distribution of photometric stellar masses, derived from SED fitting, of the passive member galaxies of XMMU J2235.3-2257. The masses of the four core early-type galaxies are shown by the dashed red histogram. Right: star formation weighted age as a function of the photometric stellar mass, both from SED fitting, of the passive cluster members. The mean mass of the XMM 2235 sample and star formation weighted age from its composite spectrophotometric data are shown by the red symbol. The best fit correlation is represented by the black line.

The parameters of this grid are the same as those of the one used in Section 6.1, with the difference that we only considered values of T up to 4.5 Gyr, which corresponds to about the age of the Universe at $z = 1.39$ assuming the standard cosmology. In Fig. 6.13, we show the distribution of the photometric stellar masses of the galaxies in the XMM 2235 sample and the star formation weighted age from the SED fit as a function of the stellar mass. As with XMMU J1229+0151, we see a clear downsizing effect.

As in Section 6.1 and in the previous Chapters, we coadded the spectra and SEDs of the galaxies in our sample and compared them to the grid of τ-models. Because we extended the considered range to wavelengths shorter than 3200 Å (see below), we computed the grid of τ-models from solar metallicity BC03 templates that used the Pickles ([1998]; also, see Chapter 2) stellar library in the mid-UV. The mean S/N, at 4100 Å rest-frame, of the individual spectra is ~3 while the composite spectrum has a S/N of 7. As in Section 6.1 we ignored the wavelength range around the telluric line at ~ 7600 Å. The characteristic dip in flux due to the undersubtraction of this line can be seen in Fig. 6.15. At $z = 1.39$, the usual wavelength range around the 4000 Å break that we used previously is at the red edge of the spectrum, with Hδ being the longest wavelength feature visible. Consequently, the region around the 4000 Å is somewhat affected by noise. We therefore expanded the wavelength range to consider mid-UV spectral features such as MgII and MgI, at 2800 and 2852 Å respectively, and the associated spectral break at 2900 Å. The 2900 Å break correlates in stars with spectral type (Fanelli et al. [1992]), but is only mildly metallicity sensitive, while the Mg II feature is most pronounced in F-type stars. These features thus

6.2 The cluster XMMU J2235.3-2557 at $z = 1.39$

make a suitable tracer of past star formation and have been used to determine the ages of high redshift galaxies (e.g. McCarthy et al. [2004a]). In Fig. 6.14, we show the confidence regions of the fit to the composite SED and spectrum of the XMM 2235 sample and the average SED with best fitting models. In Fig. 6.15, we show the composite spectrum of the XMM 2235 sample with the best fit model to the spectrophotometric data. As a comparison, we also plot the composite spectrum of field galaxies in the redshift range $1.3 < z < 1.5$ taken from the GMASS (Kurk et al. [2008]) and GDDS (Abraham et al. [2004]) surveys. We see that the 4000 Å break is deeper in the XMM 2235 spectrum, suggesting that the age difference observed at $z \sim 1.2$ (see Chapter 4) between field and cluster galaxies is also present in some form at $z \sim 1.4$. For the best fitting models, we found a star formation weighted age of $2.6^{+0.5}_{-0.5}$ Gyr, corresponding to a formation redshift of $3.3^{+1.2}_{-0.6}$, and a look-back time from $z = 1.39$ to the final formation time of $1.6^{+0.6}_{-0.4}$ Gyr, corresponding to a final formation redshift of $2.2^{+0.1}_{-0.2}$.

Clusters at low to intermediate redshifts are well known to display a segregation in their stellar population properties, with older more massive galaxies residing in the core. It is interesting to investigate whether at these large lookback times (i.e. closer to the formation redshift of galaxies) such a segregation is already in place. To this end, we divided the XMM 2235 sample into a "core" subsample, comprising the four central galaxies (within a radius of 12"~ 100 kpc, see Fig. 6.12), and those other 11 galaxies. We found a stark difference between the average stellar population of the core galaxies and of the rest, as shown in Fig. 6.16). The former appear very old, with a star formation weighted age of $4.0^{+0.2}_{-0.2}$ Gyr and a lookback time to the final formation of $3.9^{+0.3}_{+0.4}$ Gyr, consistent with a single burst of star formation at high (> 5) redshift. The other sample, outside the core, shows strong post-starburst/post-star-forming features not seen in the individual, low-S/N spectra. This can be appreciated in Fig. 6.16, where we plot the composite spectra of the two sub-samples, as well as the best fit models to the spectrophotometry. For the external early-type galaxies, the mean ages of best fitting models are $1.1^{+0.1}_{-0.1}$ Gyr for the star formation weighted age and $0.7^{+0.2}_{-0.1}$ Gyr for the final formation lookback time, corresponding to redshifts of $1.9^{+0.04}_{-0.1}$ and $1.7^{+0.1}_{-0.1}$ respectively. As found in Chapter 5, this shows that the quenching of star formation in galaxies occurs at the outskirts of the cluster, as they fall in towards the center. The evidence that such a strong gradient in the ages of early-type galaxies is already in place at $z = 1.4$ suggests that we are indeed approaching the formation epoch of the early-type galaxy population.

It is also of interest to compare the average ages of the whole cluster early-type galaxy population with the other clusters in this work. We see that the formation epoch of XMM 2235 is about the same than that of the bright red sequence sample of RX J0152.7-1357 (i.e. bin (1), see Chapter 5) and ~ 0.3 Gyr older than that of the XMM 1229 sample but ~ 0.4 Gyr lower than the formation epoch of the RDCS 1252 sample (see Chapter 4, also Fig. 6.18). This latter difference can not be ascribed to a "progenitor bias". Indeed, because at each redshift we do not sample all the progenitors of later early-type galaxies but only the oldest ones (i.e. those that are already passive), the apparent formation epochs of the same galaxy population observed at two different epochs can only decrease with redshift. This means that, at the cosmic epoch at which we observe XMMU J2235.3-2257,

6. Moving to higher redshifts: two clusters at $z \sim 1$ and $z \sim 1.4$

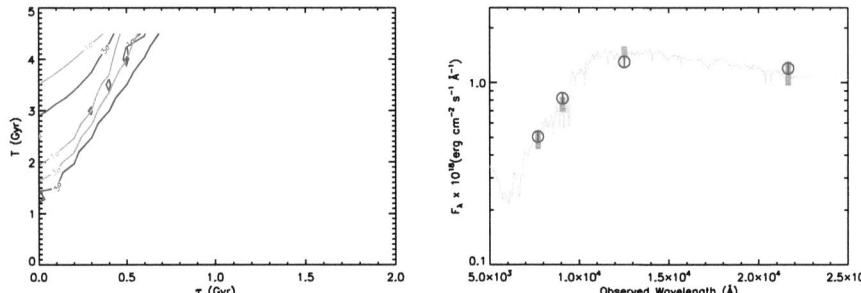

Figure 6.14: Left: 1σ and 3σ confidence regions of the fit to the average SED (blue) and spectrum (red) of the passive galaxies in the XMM 2235 sample. Right: average SED of the passive galaxies in the XMM 2235 sample (blue) and best fitting models (red) within the 3σ confidence regions of the fit. The best fit model to the average SED is shown in green (right).

the cluster RDCS J1252.9-2927 was already more evolved. This difference in formation epochs between clusters can be used to test models of galaxy formation and evolution. In Fig. 6.17, we plot the evolution between $z = 0.7$ and $z = 1.5$ of the mean rest-frame $U - V$ color of the best fitting models to the four cluster samples, compared with the scatter in color of simulated cluster galaxies from the models of Menci et al. ([2008]) described in Chapter 4. Because of the discrepancy between the composite SED and spectrum of the XMM 1229 sample, we also plot for this cluster the mean color of best fitting models to the composite SED only. We found that the semi-analytic models reproduce well the observed cluster-to-cluster color variance. This was noted as well in Menci et al. ([2008]), where the predictions of the model were compared with HST/ACS observations of a sample of eight clusters (among which RX J0152.7-1357 and RDCS J1252.9-2927) in the same redshift range.

6.2 The cluster XMMU J2235.3-2557 at $z = 1.39$

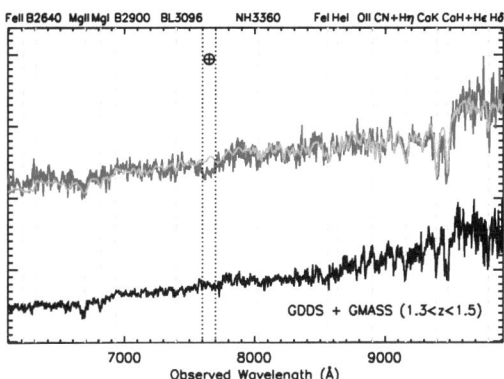

Figure 6.15: Average spectrum of the passive galaxies in the XMM 2235 sample (blue) with best fit model spectrum (red). Relevant spectral features are shown by dashed green lines and the telluric line at ~ 7600 Å by dotted lines. For comparison, the composite spectrum of field early-type galaxies from the GDDS and GMASS catalogs in the redshift range $1.3 < z < 1.5$ is shown in black.

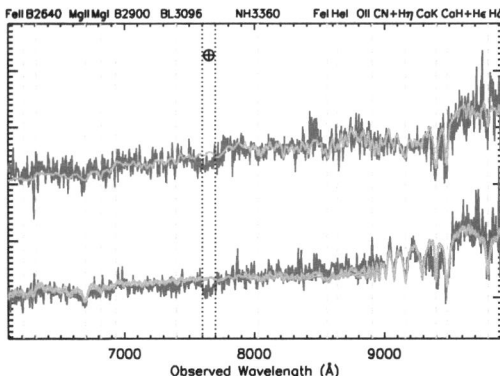

Figure 6.16: Composite spectrum of the four central passive galaxies (top) and of the other 11 galaxies in the XMM 2235 sample outside the core (bottom), in blue. The spectra of the best fit models to the spectrophotometry are shown in red. Relevant spectral features are shown by dashed green lines and the telluric line at ~ 7600 Å by dotted lines.

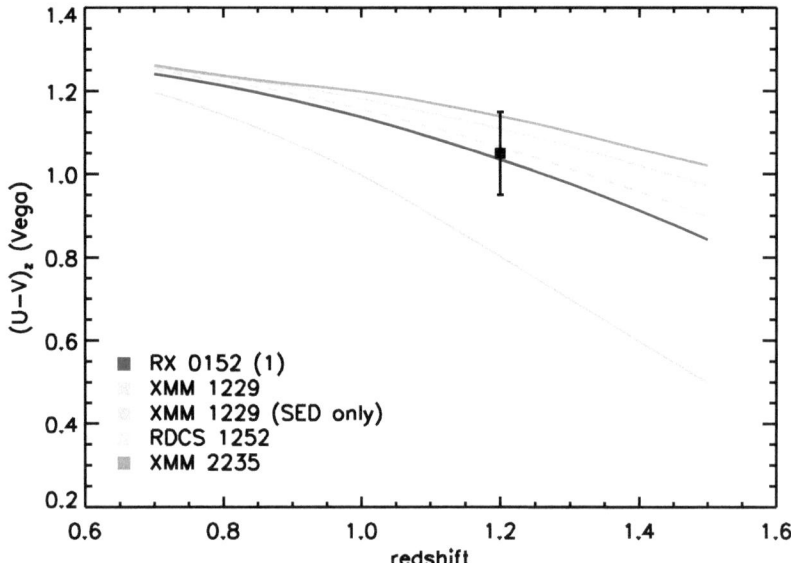

Figure 6.17: Mean rest-frame $U - V$ color (in the Vega system), as a function of redshift, of best fitting models to the average spectrophotometric data of the RX 0152 (bright red sequence), XMM 1229, RDCS 1252 and XMM 2235 samples. Because of the discrepancy between the average spectrum and SED of XMM 1229, we also plot the best fitting models to the average SED only of XMM 1229 (dashed line). The square symbol with error bars shows the mean color and 1σ spread of model cluster galaxies at $z = 1.2$ from Menci et al. ([2008]).

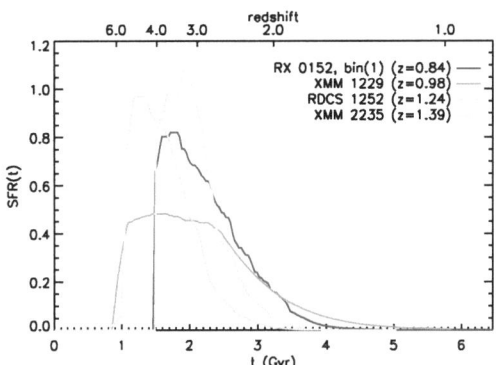

Figure 6.18: Star formation histories of the four cluster samples used in this work: median star formation histories of the best fitting models to the spectrophotometric data of the RX 0152 (bright end of the red sequence), XMM 1229, RDCS 1252 and XMM 2235 samples of cluster early-type galaxies, in blue, red, green and orange respectively. The lower x-axis shows the cosmic time t and the upper x-axis the corresponding redshift.

6.3 Summary

In this Chapter, we have used two samples of 15 and 8 early-type galaxies to investigate the histories of star formation in the two X-ray selected clusters XMMU J2235.3-2257 at $z \sim 1.4$ and XMMU J1229+0151 at $z \sim 1$ respectively. Firstly, we derived photometric stellar masses for the galaxies in each cluster. We found a correlation between the star formation weighted ages t_{SFR} of the best fit models to the SEDs of individual galaxies and their photometric stellar masses $M^\star_{phot}$, the well-known "downsizing" effect. As in Chapter 5, this is likely mostly due to an actual age difference, with only a minor contribution by metallicity. Indeed, the age difference (a few Gyr) between the highest and lowest mass galaxies in our sample can not be reproduced assuming the metallicity range observed in low and intermediate redshift galaxies (e.g. Thomas et al. [2005], Gallazzi et al. [2006]; also, see Chapter 2). As the slope of the red sequence does not change significantly out to $z \sim 1.4$ (e.g. Blakeslee et al. [2003], Mei et al. [2006b], Santos et al. [2009], Lidman et al. [2008]), and is widely interpreted as being due to a metallicity variation along the red sequence, we can safely assume the same metallicity range at high redshift. We compared the t_{SFR}-$M^\star_{phot}$ relation of the two samples to that of the early-type galaxies in the RDCS J1252.9-2927 and RX J0152.7-1357 clusters (see Fig. 6.10). Interestingly, the best fit correlation lines are very similar in slope, which suggests that star formation is quenched in cluster galaxies at roughly the same environmental density across clusters.

Secondly, we modeled the composite 4-band SED and spectrum of the passive spectroscopic members of each cluster. A third of the passive spectroscopic members of XMMU J1229+0151 is of the post-starburst/post-star-forming type and, as in RX J0152.7-1357, avoid the cluster center. We excluded these post-starburst/post-star-forming galaxies from the sample of XMMU J1229+0151 early-types to avoid biasing it. From the star formation weighted ages of the best fitting models to the spectrophotometric data of both samples, we derived a mean formation redshift of 3.1 for XMM 1229 and 3.3 for XMM 2235. Both are consistent with that of the early-type galaxies in RX J0152.7-1357 but lower than that of the RDCS 1252 sample ($z_f \sim 4$), as shown in Fig. 6.18. In the case of XMM 1229 and RX 0152, this can be ascribed to progenitor bias, as at $z < 1$ we are likely to include in our sample galaxies which would have been still active at $z \sim 1.2$. XMM 2235, however, is at a higher redshift than RDCS 1252 and so should display an earlier formation epoch had the two clusters comparable galaxy populations. The different mean formation redshifts are therefore not an artifact of the relative epochs at which the clusters are observed, but rather the evidence of different assembly histories. This variance in the formation epochs of cluster galaxies can provide a further test for models of galaxy formation. So far, the models seem to reproduce it very well but a more systematic study of high redshift ($z \gtrsim 1.4$) clusters is needed in order to quantify it precisely. We also found that the highest redshift cluster in our study shows a strong radial gradient in the mean ages of early-type galaxies. We interpret this as the evidence that, by furthering these studies at large lookback times (9 Gyr for XMMU J2235.3-2257), one can readily observe the building of the old galaxy population from the core outwards.

Chapter 7
Summary

In this thesis, we have studied several aspects of the evolution of high-redshift ($0.8 < z < 1.4$) early-type galaxies across a range of environments, by modeling their stellar population properties to infer their star formation histories. For this purpose, we have used an exceptional dataset, in terms of quality, depth and wavelength coverage, combining spectro-photometric observations from the VLT, HST, and Spitzer telescopes, which is hardly matched by other investigations. The main results of this work are the following:

- We have developed a **novel method** which **combines both the spectral energy distributions (SED) and spectra** of galaxies to model the underlying stellar populations with spectral synthesis models. Similar studies in the literature are solely based either on SED fitting or on spectral indices modeling and, to our knowledge, no attempt had been made so far to take advantage of both spectroscopic and photometric data in modeling the stellar populations of distant early-type galaxies. This allowed us to obtain stronger constraints on the star formation histories, by also mitigating inherent degeneracies among stellar population parameters.

- A principal parameter obtained by SED modeling is the stellar mass (the so-called "photometric mass"), which is widely used in current surveys to characterize galaxy evolution. In order to investigate the robustness of our photometric stellar masses, we have used a sample of low-to-intermediate redshift lensing elliptical galaxies from the literature (the so-called SLACS sample), where the stellar mass was independently estimated using a joint strong lensing and dynamical analysis. We used SDSS photometry to model galaxy SEDs and found an **excellent agreement between these independent stellar mass estimates**, specifically when a Salpeter initial mass function and solar metallicity is assumed in the models (Grillo, Gobat et al. [2008]).
We also found a clear **proportionality between the *stellar* and *total* masses**. The differences in lens geometry within the galaxy sample allowed us to study the amount and distribution of dark matter in these elliptical galaxies, within one effective radius of the center. We found that a large amount ($\gtrsim 30\%$) of dark matter is present in the central regions of the lens galaxies, consistently with previous studies and that the **profile of the**

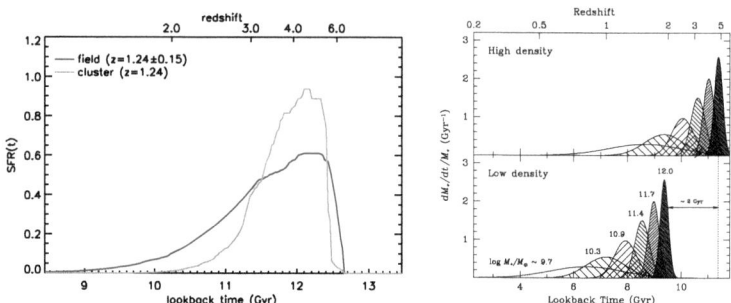

Figure 7.1: Star formation histories in low and high density environments: median star formation history of early-type galaxies in the cluster and field with stellar masses higher than $5 \times 10^{10} M_\odot$, from the spectrophotometric analysis of the GOODS field and RDCS 1252 cluster samples (left, see Chapter 4), compared with the star formation histories of early-type galaxies in the nearby universe (right) proposed by Thomas et al. ([2005]).

dark matter distribution is similar to (or even shallower than) **that of the luminous (stellar) matter**, up to one effective radius from the center. With future large samples of lensing galaxies (e.g. those expected from the Pan-STARRS survey), it will be possible to have a detailed view of the dark/luminous matter distribution with this method, particularly taking advantage of systems with multiple lensed sources (e.g. Gavazzi et al. [2008]).

- We carried out a crucial test for galaxy formation models by comparing the star formation histories of field and cluster galaxies at the highest redshift currently accessible. For this test, we used multi-wavelength, multi-observatory data for the massive cluster RDCS J1252.9-2927 at $z = 1.24$ and an homogeneous dataset from the GOODS project for the low-density environment (Gobat et al. [2008]). We found a **difference of ~ 0.5 Gyr between** the star formation timescales of early-type galaxies in **field and cluster** environments at $z \sim 1.2$. Star formation in the field galaxies is not delayed with respect to contemporary cluster galaxies (i.e. it starts approximately at the same time) but is rather more protracted, as illustrated in Fig. 7.1. This result strengthens the evidence found in some independent studies at intermediate to high redshift, based on the evolution of the fundamental plane, but appears to be in contrast with the large difference ($\gtrsim 1.5$ Gyr) observed at low redshift from fossil record data (see Fig. 7.1, right panel). We conclude that massive early-type galaxies in both high and low-density environments start forming stars at $z \sim 4 - 5$. Because the star formation processes are accelerated by the environment, cluster galaxies are the first to stop and become passive at $z \sim 2$. We would therefore expect the age difference between field and cluster galaxies of a given mass to increase with time until all field early-type galaxies in that mass range are formed. In this scenario, the evolution of the age difference between cluster and field early-type population, from

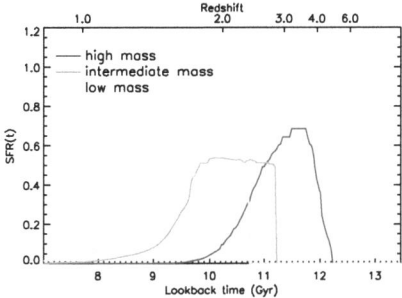 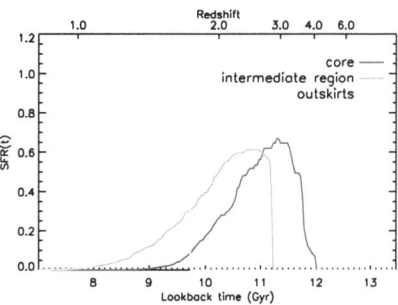

Figure 7.2: Star formation histories in the $z = 0.84$ cluster RX J0152.7-1357: median star formation history of early-type galaxies as a function of mass (left) and clustercentric distance (right, see Chapter 5).

~ 0.5 Gyr at $z \gtrsim 1$ to ~ 1.5 Gyr at low redshift, can be reproduced assuming that a non negligible fraction of the stellar mass ($\sim 10\%$) is assembled at $z < 1$. As the number of massive ($M_\star > 10^{11} M_\odot$) early-type galaxies has been found to be nearly constant since $z \sim 1$ (e.g. Cimatti et al. [2006], Faber et al. [2007]), this late star formation would have been limited to the low mass galaxy population, in agreement with the "downsizing" paradigm.

From the modeling of the photometric properties of red sequence galaxies in the $z = 1.24$ cluster RDCS J1252.9-2927, we also conclude that the tight **red sequence** ($\sigma \simeq 0.05$ mag) observed in clusters at $1 < z < 1.4$ was **established over ~ 1 Gyr**, starting at $z \sim 2$. This prediction is consistent with studies of forming red sequences in protoclusters at $z > 2$, which find a large scatter.

Star formation histories from this study were also compared in detail with the predictions of state-of-the-art semi-analytic models of galaxy evolution (Menci, Rosati, Gobat et al. [2008]).

• We have studied the stellar population properties of early-type galaxies in the rich, massive cluster RX J0152.7-1357 at $z = 0.837$ and their dependence on environment and on intrinsic galaxy properties (e.g. stellar mass), taking advantage of a large sample of spectroscopic members. We found that cluster galaxies were strongly (but not completely) segregated in mass, with the most massive ones occupying the core of the cluster. From spectrophotometric modeling, we found a **strong radial age gradient**, with the central galaxies being also the oldest (see Fig. 7.2). The core early-type galaxies are old (~ 4.5 Gyr), having formed at $z > 3$ and stopped star formation altogether at $z \sim 2$, while early-type galaxies in the cluster outskirts are ~ 1.5 Gyr younger. This result paints a scenario where the central galaxies formed rapidly at high redshift and ceased forming stars early while galaxies outside the immediate core continued star formation for at least 1 Gyr longer (Demarco, Gobat et al. [2009]). This **environmental age gradient** mirrors the difference we found between the field and cluster early-type galaxy population, but is steeper, and shows that galaxy evolution is strongly driven by environment, in agreement with the hierarchical formation scenario.

Furthermore, we found a population of **post-starburst/post-star-forming galaxies**, which became passive less than 1 Gyr ago. They occupy the **faint blue end of the cluster red sequence** and lie **outside the densest cluster regions**. This implies that star formation is suppressed in cluster galaxies well before they reach the center. Interestingly, most of those post-starburst/post-star-forming galaxies lie outside the region of strong X-ray emission as well. This would suggests that these galaxies did not lose their gas through direct interaction with the dense intracluster medium (ICM) and therefore would exclude ram pressure (Gunn & Gott [1972]) and shock heating as a mechanisms for the depletion of gas in these galaxies. Other processes include gravitational interaction with the cluster potential, which would strip the hot gas reservoir of a galaxy (e.g. Larson et al. [1980]), or interaction with other cluster galaxies, which would trigger gas-consuming starbursts.

- We have then extended our spectrophotometric modeling method to two other high-redshift X-ray selected clusters, XMMU J1229+0151 at $z = 0.98$ (Santos, Rosati, Gobat et al. [2009]) and XMMU J2235.3-2257 at $z = 1.39$, the second most distant X-ray luminous cluster known to date. From SED modeling, we derived ages and masses for the early-type galaxies in both clusters and compared them to the ages and masses of the early-type galaxies in the first two clusters, at $z = 0.84$ and $z = 1.24$. For all clusters, we found **relations between age and stellar mass** with **very similar slopes**. This implies that the star formation timescales of early-type galaxies are similar in all four clusters and suggests that star formation in early-type galaxies ceases in those clusters through the same mechanism.

Through spectrophotometric analysis, we estimated that the massive early-type galaxies in both clusters have formed at $z \gtrsim 3$, approximately 0.5 Gyr later than the cluster galaxies at $z = 1.24$. While it is not surprising that a sample of early-type galaxies at a lower redshift might appear younger, as we might include in the lower redshift sample galaxies which would not have been passive at $z = 1.24$ (the so-called progenitor bias) the same effect would also make the higher redshift sample appear older. We conclude then that the observed difference between the early-type galaxies of RDCS J1252.9-2927 at $z = 1.24$ and XMMU J2235.3-2257 at $z = 1.39$ is the result of distinct formation histories. This **variance in the formation epochs** of galaxy clusters thus provides a further interesting test for the current models of galaxy formation.

While the central galaxies of both clusters are old, we also found in both cases a population of post-starburst/post-star-forming galaxies which tend to avoid the cluster cores and, in the case of XMMU J2235.3-2257 at $z = 1.39$, have only very recently become passive. These **strong radial age gradients**, which seem to increase with redshift, show that we are approaching, at $z \sim 1.4$, the formation epoch of the early-type cluster population (see Fig. 7.3).

In light of this, several developments can be considered:

- the redshift range $1 < z < 2$ has been called the "cluster desert", as very few galaxy clusters have been found in this range and none at $z > 1.5$ so far. This is mostly due to the low density of high-redshift clusters on the sky, making them hard to find in X-ray or optical/near-IR surveys. However, a number of new programs specifically aimed at detecting high-redshift clusters have been undertaken and ongoing or upcoming surveys, notably those making use of the Sunyaev Zel'dovich effect (S-Z), promise to increase the number of high-redshift galaxy clusters by an order of magnitude at least. This depends however on the actual space density of massive clusters at $z > 1$, as the best S-Z methods are currently sensitive to clusters with masses exceeding $\sim 2 \times 10^{14} M_\odot$. A detailed study of clusters found at $z > 1.4$ would bridge the gap between the protoclusters that have been found at $z > 2$ (e.g. Kurk et al. [2001]), with strongly star forming structures surrounding a massive radio galaxy, and the $z < 1.4$ clusters studied here, which have a well-defined passive galaxy population. In particular, the strong age segregation observed in XMMU J2235.3-2257 suggests that, in this higher redshift range, one would observe the building of the red sequence population and thus shed light on the processes that drive it.

- on the other hand, post-starburst galaxies in the clusters already known at $z \sim 1$ can also provide precious insight on the environmental processes at work in these high-redshift clusters. Post-starburst galaxies in the nearby Universe have been found to be predominantly the result of galaxy-galaxy interactions (e.g. Zabludoff et al. [1996], Goto [2005]). While the results of this work suggest that post-starburst galaxies in high-redshift clusters also were not quenched through interaction with the dense intracluster medium, the mechanism behind the quenching of star formation is not clear. A morphological or a dynamical study of these post-starburst galaxies, using AO-assisted instruments such as SINFONI on the VLT (Thatte et al. [1998]), would be important in determining the physical process that led to the cessation of star formation, whether due to interaction with the cluster potential or with other cluster galaxies, and its timescale.

- as with clusters, field galaxies at $z > 1.5$ are especially interesting in the context of galaxy formation. From our comparison of field and cluster galaxies at $z \sim 1.2$, we expect the progenitors of lower redshift field early-type galaxies to evolve rapidly between $z \sim 2$ to $z = 1.5$, changing from actively star forming systems to passive ones. The identification and characterization of the earliest field early-type galaxies, in particular, would provide a very strong test for formation models. The $z \lesssim 2$ redshift range has recently been probed by surveys such as GMASS and GDDS and will be explored more fully by collaborations such as zCOSMOS (Lilly et al. [2007]). At $z > 2$, one finds mostly actively star forming galaxies and protogalaxies. The cosmic star formation rate at high redshift, as well as the high metal abundance measured in the ICM of clusters at $z > 1$ (e.g. Balestra et al. [2007]), provides an

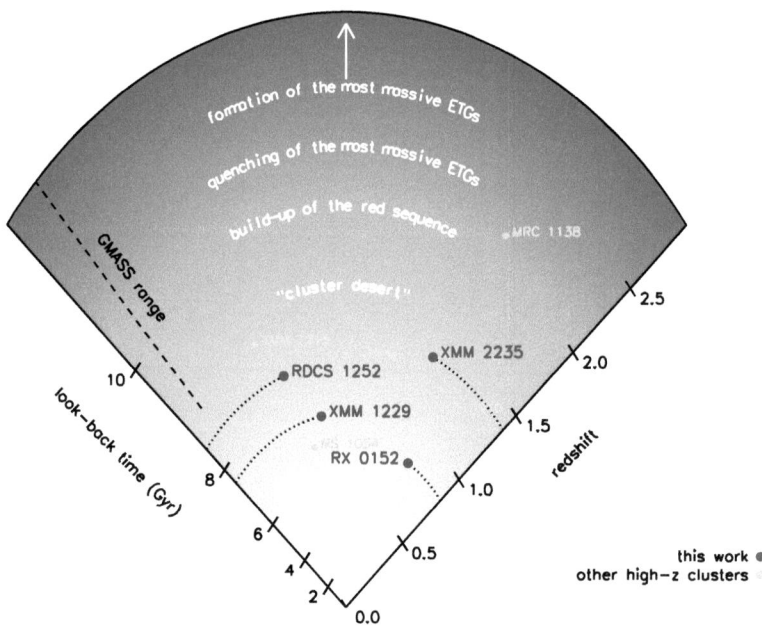

Figure 7.3: Timeline of early-type galaxy formation and cluster assembly. The clusters used in this work are represented by blue circles while other high-redshift X-ray selected clusters, as well as the protocluster MRC 1138-262 at $z = 2.16$ (Kurk et al. [2001]), are shown in green. The redshift range of the GMASS (Kurk et al. [2008]) spectroscopy is also shown.

additional constraint on galaxy formation models and the processes which lead to the formation and segregation of baryons in their hot and cold phases.

Bibliography

[2004] Abraham, R.G., Glazebrook, K., McCarthy, P.J. et al. 2004, AJ, 127, 2455

[2006] Adelman-McCarthy, J.K., Agüeros, M.A., Allam, S.S. et al. 2006, ApJS, 162, 38

[1993] Alongi, M., Bertelli, G., Bressan, A. et al. 1993, A&AS. 97, 851

[1992] Appenzeller, I., Rupprecht, G. 1992, ESO Messenger, 67, 18

[1987] Arimoto, N., Yoshi, Y. 1987, A&A, 173, 23

[1976] Avni, Y. 1976, ApJ, 210, 642

[2007] Balestra, I., Tozzi, P., Ettori, S. et al. 2007, A&A, 462, 429

[2004] Baldry, I.K., Glazebrook, K., Brinkmann, J. et al. 2004, ApJ, 600, 681

[1999] Balogh, M.L., Morris, S.L., Yee, H.K.C. et al. 1999, ApJ, 527, 54

[1959] Baum, W.A. 1959, IAU Symp., 10, 23

[2001] Bell, E.F., de Jong, R.S. 2001, ApJ, 550, 212

[2003] Bell, E.F., McIntosh, D.H., Katz, N., Weinberg, M.D. 2003, ApJS, 149, 289

[2000] Benítez, N. 2000, ApJ, 536, 571

[1998] Bernardi, M., Renzini, A., Da Costa, L.N. et al. 1998, ApJ, 507, 43

[2005] Bernardi, M., Sheth, R.K., Nichol, R.C., Schneider, D.P., Brinkmann, J. 2005, AJ, 129, 61

[1996] Bertin, E., Arnouts, S. 1996, A&AS, 117, 393

[1987] Binney, J., Tremaine, S. 1987, Galactic Dynamics, Princeton University Press

[2003] Blakeslee, J.P., Franx, M., Postman, et al. 2003, ApJ, 596, L143

[2006] Blakeslee, J.P., Holden, B.P., Franx, M. et al. 2006, ApJ, 644, 30

[2005] Böhringer, H., Mullis, C., Rosati, P. et al. 2005, ESO Messenger, 120, 33

[2004] Bolton, A.S., Burles, S., Schlegel, D.J., Eisenstein, D.J., Brinkmann, J. 2004, AJ, 127, 1860

[2006] Bolton, A.S., Burles, S., Koopmans, L.V.E., Treu, T., Mousakas, L.A. 2006, ApJ, 638, 703

[1991] Bond, J.R., Cole, S., Efstathiou, G., Kaiser, N. 1991, ApJ, 379, 440

[1992] Bower, R.G., Lucey, J.R., Ellis, R.S. 1992, MNRAS, 254, 601

[1999] Bower, R.G., Terlevich, A., Kodama, T., Caldwell, N. 1999, ASP Conf. Ser., 163, 211

[2004] Bradač, M., Lombardi, M., Schneider, P. 2004, A&A, 424, 13

[1993] Bressan, A., Fagotto, F., Bertelli, G., Chiosi, C. 1993, A&AS, 100, 647

[2003] Bruzual, G., Charlot, S. 2003, MNRAS, 344, 1000

[2006] Bundy, K., Ellis, R.S., Conselice, C.J. et al. 2006, ApJ, 651, 120

[2000] Calzetti, D., Armus, L., Bohlin, R.C. et al. 2000, ApJ, 533, 682

[1989] Cardelli, J.A., Clayton, G.C., Mathis, J.S. 1989, ApJ, 345, 245

[2002] Carlstrom, J.E., Holder, G.P., Reese, E.D. 2002, ARA&A, 40, 643

[2008] Cassata, P., Cimatti, A., Kurk, J. et al. 2008, A&A, 483, 39

[1997a] Cassisi, S., Castellani, M., Castellani, V. 1997, A&A, 317, 108

[1997b] Cassisi, S., degl'Innocenti, S., Salaris, M. 1997, MNRAS, 290, 515

[2000] Cassisi, S., Castellani, V., Ciarcelluti, P., Piotto, G., Zoccali, M. 2000, MNRAS, 315, 679

[2008] Cattaneo, A., Dekel, A., Faber, S.M., Guiderdoni, B. 2008, MNRAS, 389, 567

[2000] Cavaliere, A., Vittorini, V. 2000, ApJ, 543, 599

[2003] Chabrier, G. 2003, PASP, 115, 763

[2006] Cimatti, A., Daddi, E., Renzini, A. 2006, A&A, 453, 29

[2006] Clemens, M.S., Bressan, A., Nikolic, B. et al. 2006, MNRAS, 370, 702

[2005] Coia, D., McBreen, B., Metcalfe, L. et al. 2005, A&A, 431, 433

[1980] Coleman, G.D., Wu, C.-C., Weedman, D.W. 1980, ApJS, 43, 393

[2007] Cooper, M.C., Newman, J.A., Coil, A.L. et al. 2007, MNRAS, 376, 1445

[1987] Couchm W.J., Sharples, R.M. 1987, MNRAS, 229, 423

[1996] Cowie, L.L., Songaila, A., Hu, E.M., Cohen, J.G. 1996, AJ, 112, 839

[2005] Daddi, E., Renzini, A., Pirzkal, N. et al. 2005, ApJ, 626, 680

[1985] Davis, M., Efstathiou, G., Frenk, C.S., White, S.D.M. 1985, ApJ, 292, 371

[2006] De Lucia, G., Springel, V., White, S.D.M., Croton, D., Kauffmann, G. 2006, MNRAS, 366, 499

[2005] Demarco, R., Rosati, P., Lidman, C. et al. 2005, A&A, 432, 381

[2007] Demarco, R., Rosati, P., Lidman, C. et al. 2007, ApJ, 663, 164D

[2009] Demarco, R., Gobat, R. et al. 2009, in prep.

[2007] De Propris, R., Stanford, S.A., Eisenhardt, P.R., Holden, B.P., Rosati, P. 2007, AJ, 133, 2209

[1948] de Vaucouleurs, G. 1948, Ann. Astroph., 11, 247

[2003] Dickinson, M., Giavalisco, M. the GOODS team 2003, in The Mass of Galaxies at Low and High Redshift: Proceedings of the ESO and USM Workshop held in Venice, Italy, 24-26 October 2001. Also, astro-ph/0204213

[1987] Djorgovski, S., Davis, M. 1987, ApJ, 313, 59

[1980] Dressler, A. 1980, ApJS, 42, 565

[1987] Dressler, A., Lynden-Bell, D., Burstein, D. et al. 1987, ApJ, 313, 42

[1999] Dressler, A., Smail, I., Poggianti, B.M. et al. 1999, ApJS, 122, 51

[2004] Dressler, A., Oemler, A., Poggianti, B.M. et al., 2004, ApJ, 617, 867

[2004] Drory, N., Bender, R., Hopp, U. 2004, ApJ, 616, 103

[2000] Ebeling, H., Jones, L.R., Permlan, E. et al. 2000, ApJ, 534, 133

[1962] Eggen, O.J., Lynden-Bell, D., Sandage, A.R. 1962, ApJ, 136, 748

[2004] Eisenhardt, P.R., Stern, D., Brodwin, M. et al. 2004, ApJS, 154, 48

[2007] Eisenhardt, P.R., De Propris, R., Gonzalez, A.H. et al. 2007, ApJS, 169, 225

[2001] Eisenstein, D.J., Annis, J., Gunn, J.E. et al. 2001, AJ, 122, 2267

[2000] Ellis, R.S., Abraham, R.G., Brinchmann, J., Menanteau, F. 2000, A&G, 41, 10

[1973] Faber, S.M. 1973, ApJ, 179, 731

[1976] Faber, S.M., Jackson, R.E. 1976, ApJ, 204, 668

[2007] Faber, S.M., Willmer, C.N.A., Wolf, C. et al. 2007, ApJ, 665, 265

[1994a] Fagotto, F., Bressan, A., Bertelli, G., Chiosi, C. 1994, A&AS, 104, 365

[1994b] Fagotto, F., Bressan, A., Bertelli, G., Chiosi, C. 1994, A&AS, 105, 29

[1985] Falco, E.E., Gorenstein, M.V., Shapiro, I.I. 1985, ApJ, 289, L1

[1992] Fanelli, M.N., O'Connell, R.W., Burstein, D., Wu, C.-C. 1992, ApJS, 82, 197

[1998] Fasano, G., Cristiani, S. Arnouts, S., Filippi, M. 1998, AJ, 115, 1400

[2008] Fassbender, R., Böhringer, H., Santos, J. et al. 2008, in Relativistic Astrophysics and Cosmology - Einstein's Legacy, Springer-Verlag, Berlin and Heidelberg

[1998] Fazio, G.G., Hora, J.L., Willner, S.P. et al. 1998, Proc. SPIE, 3354, 114

[2005] Ferrari, C., Benoist, C., Maurogordato, S., Cappi, A., Slezak, E. 2005, A&A, 430, 19

[2005] Ferreras, I., Prasenjit, S., Williams, L.L.R. 2005, ApJ, 625, 5

[2008] Ferreras, I., Prasenjit, S., Burles, S. 2008, MNRAS, 383, 857

[1998] Ford, H.C., Bartko, F., Bely, P.Y. et al. 1998, Proc. SPIE, 3356, 234

[1990] Frogel, J.A., Mould, J., Blanco, V.M. 1990, ApJ, 352, 96

[2006] Gallazzi, A., Charlot, S., Brinchmann, J., White, S.D.M. 2006, MNRAS, 370, 1106

[1996] Gavazzi, G., Pierini, D., Boselli, A. 1996, A&A, 312, 397

[2002] Gavazzi, G., Bonfanti, C., Sanvito, G., Boselli, A., Scodeggio, M. 2002, ApJ, 576, 135

[2007] Gavazzi, R., Treu, T., Rhodes, J.D. et al. 2007, ApJ, 667, 176

[2008] Gavazzi, R., Treu, T., Koopmans, L.V.E. et al. 2008, ApJ, 677, 1046

[2004] Giavalisco, M., Ferguson, H.C., Koekemoer, A.M. et al. 2004, ApJ, 600, 93

[1996] Girardi, L., Bressan, A., Chiosi, C., Bertelli, G., Nasi, E. 1996, A&AS, 117, 113

[2000] Girardi, L., Bressan, A., Bertelli, G., Chiosi, C. 2000, A&AS, 141, 371

BIBLIOGRAPHY

[2008] Gobat, R., Rosati, P., Strazzullo, V. et al. 2008, A&A, 488, 853

[2005] Goto, T. 2006, MNRAS, 357, 937

[2001] Graham, A.W., Trujillo, I., Caon, N. 2001, AH, 122, 1707

[2008] Grillo, C., Gobat, R., Rosati, P., Lombardi, M. 2008, A&A, 477, 25

[1972] Gunn, J.E., Gott, J.R. 1972, ApJ, 176, 1

[2007] Häussler, B., McIntosh, D.H., Barden, M. et al. 2007, ApJS, 172, 615

[1990] Hernquist, L. 1990, ApJ, 356, 359

[2004] Holden, B.P., Stanford, S.A., Eisenhardt, P., Dickinson, M. 2004, AJ, 127, 2484

[2006] Holden, B.P., van der Wel, A., Franx, M. et al. 2005, ApJ, 620, 83

[2007] Holden, B.P, Illingworth, G.D., Franx, M. et al. 2007, ApJ, 670, 190

[1926] Hubble, E.P. 1926, ApJ, 64, 321

[1919] Jeans, J.H. 1919, Problems of cosmogony and stellar dynamics, University press, Cambridge

[2005] Jee, M.J., White, R.L., Benítez, N. et al. 2005, ApJ, 618, 46

[2008] Jimenez, R., Bernardi, M., Haiman, Z., Panter, B., Heavens, A.F. 2008, ApJ, 669, 947

[1996] Jørgensen, I., Franx, M., Kjærgaard, P. 1996, MNRAS, 280, 167

[1998] Kauffmann, G., Charlot, S. 1998, MNRAS, 297, 23

[2003] Kauffmann, G., Heckman, T.M., White, S.D.M. et al. 2003, MNRAS, 341, 33

[1992] Kennicutt, R.C. 1992, ApJ, 388, 310

[1998] Kennicutt, R.C. 1998, ARA&A, 36, 189

[1997] Kodama, T., Arimoto, N. 1997, A&A, 210, 41

[1998] Kodama, T., Arimoto, N., Barger, A., Aragón-Salamanca, A. 1998, A&A, 334, 99

[2007] Kodama, T., Tanaka, I., Kajisawa, M. et al. 2007, MNRAS, 377, 1717

[2006] Koopmans, L.V.E., Treu, T., Bolton, A.S., Burles., S., Moustakas, L.A. 2006, ApJ, 649, 599

[1977] Kormendy, J. 1977, ApJ, 218, 333

[1989] Kormendy, J., Djorgovski, S. 1989, ARA&A, 27, 235

[2001] Kroupa, P. 2001, MNRAS, 322, 231

[2001] Kurk, J.D., Pentericci, L., Röttgering, H.J.A., Miley, G.K., ApSSS, 277, 543

[2008] Kurk, J.D., Cimatti, A., Daddi, E. et al. 2008, ASP Conf. Ser., 381, 303; also, astro-ph/0604132

[1993] Lacey, C., Cole, S. 1993, MNRAS, 262, 627

[2005] Lapi, A., Cavaliere, A., Menci, N. 2005, ApJ, 619, 60

[1975] Larson, R.B. 1975, MNRAS, 173, 671

[1980] Larson, R.B., Tinsley, B.M., Caldwell, C.N. 1980, ApJ, 237, 692

[2003] Le Borgne, J.F., Bruzual, G., Pelló, R. et al. 2003, A&A, 402, 433

[2003] Le Fèvre, O., Saisse, M., Mancini, D. et al. 2003, Proc. SPIE, 4841, 1670

[1997] Lejeune, T., Cuisinier F., Buser R. 1997, A&AS, 125, 229

[1998] Lejeune, T., Cuisinier F., Buser R. 1998, A&AS, 130, 65

[2004] Lidman, C., Rosati, P., Demarco, R. et al. 2004, A&A, 416, 829

[2008] Lidman, C., Rosati, P., Tanaka, M. et al. 2008, A&A, 489, 981

[2007] Lilly, S.J:, Le Fèvre, O., Renzini, A. et al. 2007, ApJS, 172, 70

[2006] Lintott, C.J., Ferreras, I., Lahav, O. 2006, ApJ, 648, 826

[1999] Loewenstein, M., White, R.E., ApJ, 518, 50

[2007] Luo, Z.-J., Shu, C.-G., Huang, J.-S. 2007, PASJ, 59, 541

[2005] Maraston, C. 2005, MNRAS, 362, 799

[2007] Marcillac, D., Rigby, J.R., Rieke, G.H., Kelly, D.M. 2007, ApJ, 654, 825

[2003] Marri, S., White, S.D.M. 2003, MNRAS, 345, 561

[2003] Maughan, B.J., Jones, L.R., Ebeling, H. et al. 2003, ApJ, 587, 589

[2004a] McCarthy, P.J., Le Borgne, D., Crampton, D. et al. 2004, ApJ, 614, 9

[2004b] McCarthy, P.J. 2004, ARA&A, 42, 477

[2006a] Mei, S., Blakeslee, J.P., Stanford, S.A. et al. 2006, ApJ, 639, 81

[2006b] Mei, S., Holden, B.P., Blakeslee, J.P. et al. 2006, ApJ, 644, 759

[2002] Menci, N., Cavaliere, A., Fontana, A: et al. 2002, ApJ, 575, 18

[2004] Menci, N., Cavaliere, A., Fontana, A. et al. 2004, ApJ, 604, 12

[2005] Menci, N., Fontana, A., Giallongo, E., Salimbeni, S. 2005, ApJ, 632, 49

[2008] Menci, N., Rosati, P., Gobat, R. et al. 2008, ApJ, 685, 863

[1992] Moorwood, A. 1992, ESO Messenger, 70, 10

[1998] Moorwood, A., Cuby, J.-G., Lidman, C. 1998, ESO Messenger, 91, 9

[2005] Mullis, C.R., Rosati, P., Lamer, G. et al. 2005, ApJ, 623, 85

[1994] Mushotzky, R.F., Loewenstein, M., Awaki, H. et al. 1994, ApJ, 436, 79

[2004] Nagamine, K., Cen, R., Hernquist, L. Ostriker, J.P., Springel, V. 2004, ApJ, 610, 45

[2005] Napolitano, N.R., Capaccioli, M., Romanowsky, A.J. et al. 2005, MNRAS, 357, 691

[1974] Oke, J.B. 1974, ApJS, 27, 21

[2001] Ortwin, G., Kronawitter, A., Saglia, R.P., Bender, R. 2001, AJ, 121, 1936

[1998] Pickles, A.J. 1998, PASP, 110, 863

[1997] Poggianti, B.M., Barbaro, G. 1997, A&A, 325, 1025

[1999] Poggianti, B.M., Smail, I., Dressler, A. et al. 1999, ApJ, 518, 576

[2004] Poggianti, B.M., Bridges, T.J., Komiyama, Y. et al. 2004, ApJ, 601, 197

[2005] Postman, M., Franx, M., Cross, N.J.G. et al. 2005, ApJ, 623, 721

[2007] Pozzetti, L., Bolzonella, M., Lamareille, F. et al. 2007, A&A, 474, 443

[2001] Prugniel, Ph., Soubiran, C. 2001, A&A, 369, 1048

[1977] Rees, M.J., Ostriker, J.P. 1975, MNRAS, 179, 541

[1986] Renzini, A., Buzzoni, A. 1986, in Chiosi, C., Renzini, A., Spectral Evolution of Galaxies, Reidel, Dordrecht

[2006] Renzini, A. 2006, ARA&A, 44, 141

[2006] Rettura, A., Rosati, P., Strazzullo, V. et al. 2006, A&A, 458, 717

[2008] Rettura, A., Rosati, P., Nonino, M. et al. 2008, ApJ, in press

[2004] Rieke, G.H., Young, E.T., Engelbracht, C.W. et al. 2004, ApJS, 154, 25

[1998] Rosati, P., della Ceca, R., Norman, C., Giacconi, R. 1998, ApJ, 492, L21

[2004] Rosati, P., Tozzi, P., Ettori, S. et al. 2004, AJ, 127, 230

[2009] Rosati, P. et al., A&A, in prep.

[1985] Rose, J.A. 1985, AJ, 90, 1927

[1992] Saglia, R.P., Bertin, G., Stiavelli, M. 1992, ApJ, 384, 433

[2000] Salasnich, B., Girardi, L., Weiss, A., Chiosi, C. 2000, A&A, 361, 1023

[1955] Salpeter, E.E. 1955, ApJ, 121, 161

[2006] Sánchez-Blázquez, P., Gorgas, J., Cardiel, N., González, J.J. 2006, A&A, 457, 809

[1978] Sandage, A., Visvanathan, N. 1978, ApJ, 225, 742

[1986] Sandage, A. 1986, A&A, 161, 89

[2009] Santos, J., Rosati, P., Gobat, R. et al. 2009, A&A, submitted

[2007] Scarlata, C., Carollo, C.M., Lilly, S.J. et al. 2007, ApJS, 172, 494

[1976] Schechter, P. 1976, ApJ, 203, 297

[1998] Schlegel, D.J., Finkbeiner, D.P., Davis, M. 1998, ApJ, 500, 525

[1992] Schneider, P., Ehlers, J., Falco, E.E. 1992, Gravitational Lenses, Springer-Verlag, New York

[1995] Schneider, P., Seitz, C. 1995, A&A, 294, 411

[1963] Sérsic, J.L. 1963, Bol. Asoc. Argentina Astron., 6, 41

[1999] Smail, I., Morrision, G., Gray, M.E. et al. 1999, ApJ, 525, 609

[2007] Spergel, D.N., Bean, R., Doré, O. et al. 2007, ApJS, 170, 377

[1951] Spitzer, L., Baade, W. 1951, ApJ, 113, 413

[1998] Stanford, S.A., Eisenhardt, P.R., Dickinson, M. 1998, ApJ, 492, 461

[2008] Staniszewski, Z., Ade, P.A.R, Aird, K.A. et al. 2008, astro-ph/0810.1578

[2006] Strazzullo, V., Rosati, P., Stanford, S.A. et al. 2006, A&A, 450, 909

[2007] Tanaka, M., Tadayuki, K., Kajisawa et al. 2007, MNRAS, 377, 1206

BIBLIOGRAPHY 141

[2008] Tanaka, M., Finoguenov, A., Kodama, T. et al. 2008, A&A, 489, 571

[2005] Temi, P., Brighenti, F., Mathews, W.G. 2005, ApJ, 635, 25

[1998] Thatte, N.A., Tecza, M., Eisenhauer, F. et al. 1998, SPIE, 3353, 704

[2005] Thomas, D., Maraston, C., Bender, R., Mendes de Oliveira, C. 2005, ApJ, 621, 673

[1980] Tinsley, B.M. 1980, Fundamentals of Cosmic Physics, Vol. 5, Gordon & Breach

[2004] Toft, S., Mainieri, V., Rosati, P. et al. 2004, A&A, 422, 29

[1977] Toomre, A. 1977, Evolution of Galaxies and Stellar Populations, BM Tinsley & RB Larson, New Haven:Yale University Observatory

[2000] Trager, S.C., Faber, S.M., Worthey, G., González, J.J. 2000, AJ, 120, 165

[2001] Tran, H.D., Tsvetanov, Z., Ford, H.C., Davies, J. 2001, ApJ, 121, 2928

[2004] Tremonti, C.A., Heckman, T.M., Kauffmann, G. et al. 2004, ApJ, 613, 898

[2004] Treu, T., Koopmans, L.V.E. 2004, ApJ, 611, 739

[2005] Treu, T., Ellis, R.S., Liao, T.X., van Dokkum, P.G. 2005, ApJ, 622, 5

[2006] Treu, T., Koopmans, L.V., Bolton, A.S., Burles, S., Moustakas, L.A. 2006, ApJ, 640, 662

[2005] van der Wel, A., Franx, M., van Dokkum, et al. 2005, ApJ, 631, 145

[2006] van der Wel., A., Franx, M., Wuyts, S. et al. 2006, ApJ, 652, 97

[2000] van Dokkum, P.G., Franx, M., Fabricant, D., Illingworth, G.D., Kelson, D.D. 2000, ApJ, 541, 95

[2001a] van Dokkum, P.G., Franx, M., Kelson, D.D., Illingworth, G.D. 2001, ApJ, 553, 39

[2001b] van Dokkum, P.G., Franx, M. 2001, ApJ, 553, 90

[2003] van Dokkum, P.G., Stanford, S.A. 2003, ApJ, 585, 78

[2007] van Dokkum, P.G., van der Marel, R.P. 2007, ApJ, 655, 30

[2005] Vanzella, E., Cristiani, S., Dickinson, M. et al. 2005, A&A, 434, 53

[2006] Vanzella, E., Cristiani, S., Dickinson, M. et al. 2006, A&A, 454, 423

[2008] Vanzella, E., Cristiani, S., Dickinson, M. et al. 2008, A&A, 478, 83

[1993] Vassiliadis, E., Wood, P.R. 1993, ApJ, 413, 641

[2002] Westera, P., Lejeune, T., Buser, R., Cuisinier, F., Bruzual, G. 2002, A&A, 381, 524

[1978] White, S.D.M., Rees, M.J. 1978, MNRAS, 183, 341

[2005] Wolf, C., Gray, M.E., Meisenheimer, K. 2005, A&A, 443, 435

[1997] Worthey, G., Ottaviani, D.L. 1997, ApJS, 111, 377

[2000] York, D.G., Adelman, J., Anderson, J.E. et al. 2000, AJ, 120, 1579

[1996] Zabludoff, A.I., Zaritsky, D., Lin, H. et al. 1996, ApJ, 466, 104

[1999] Ziegler, B.L., Saglia, R.P., Bender, R. et al. 1999, A&A, 346, 13

[2007] Zirm, A.W., Stanford, S.A., Postman, M. et al. 2008, ApJ, 680, 224

Acknowledgements

I would like to express my gratitude to my thesis supervisor, Dr. Piero Rosati, for his patience and the numerous ideas, the good advice, the encouragements and overall support he provided throughout my work, as well as to Dr. Hans Böhringer for his constructive comments and suggestions, which helped me improve this thesis. I also wish to thank Claudio Grillo, Ricardo Demarco, Veronica Strazzullo, Joana Santos, Nicola Menci and Mario Nonino for the fruitful collaborations and insightful conversations. In addition, I thank Matt Lehnert for his early advice and for helping me obtain a studentship at ESO. Furthermore, I wish to thank the European Organisation for Astronomical Research in the Southern Hemisphere (ESO), and especially Bruno Leibundgut, for supporting this work and providing a stimulating work environment. I am also grateful to the Excellence Cluster Universe for financially supporting the final part of this work.

Finally, I want to thank my family for the love and attention they gave me throughout my life and without which I could probably not have completed this thesis.

Die VDM Verlagsservicegesellschaft sucht für wissenschaftliche Verlage abgeschlossene und herausragende

Dissertationen, Habilitationen, Diplomarbeiten, Master Theses, Magisterarbeiten usw.

für die kostenlose Publikation als Fachbuch.

Sie verfügen über eine Arbeit, die hohen inhaltlichen und formalen Ansprüchen genügt, und haben Interesse an einer honorarvergüteten Publikation?

Dann senden Sie bitte erste Informationen über sich und Ihre Arbeit per Email an *info@vdm-vsg.de*.

Sie erhalten kurzfristig unser Feedback!

VDM Verlagsservicegesellschaft mbH
Dudweiler Landstr. 99   Telefon +49 681 3720 174
D - 66123 Saarbrücken   Fax   +49 681 3720 1749
www.vdm-vsg.de

Die VDM Verlagsservicegesellschaft mbH vertritt

Printed by Books on Demand GmbH, Norderstedt / Germany